AF753351

Energy Storage Technologies and Applications

For a complete listing of titles in the
Artech House Power Engineering Library,
turn to the back of this book.

Energy Storage Technologies and Applications

C. MICHAEL HOFF

ARTECH HOUSE
BOSTON | LONDON
artechhouse.com

Library of Congress Cataloging-in-Publication Data
A catalog record for this book is available from the U.S. Library of Congress.

British Library Cataloguing in Publication Data
A catalogue record for this book is available from the British Library.

Cover design by Andy Meaden

ISBN 13: 978-1-63081-908-8

© 2022 Artech House
685 Canton Street
Norwood, MA 02062

All rights reserved. Printed and bound in the United States of America. No part of this book may be reproduced or utilized in any form or by any means, electronic or mechanical, including photocopying, recording, or by any information storage and retrieval system, without permission in writing from the publisher.

All terms mentioned in this book that are known to be trademarks or service marks have been appropriately capitalized. Artech House cannot attest to the accuracy of this information. Use of a term in this book should not be regarded as affecting the validity of any trademark or service mark.

10 9 8 7 6 5 4 3 2 1

Contents

Foreword

Subscribers to news updates relating to advancements in energy storage technologies are very familiar with articles describing the next "lithium-ion killer," with higher energy density, faster recharge, or improved cycle life. Other announcements relate to new technologies for long-duration storage, perhaps offering high efficiency and low capital cost. Some of these advances are genuine and are likely to be incorporated into commercial energy storage systems in the near future; others may be premature, citing laboratory test results for products that still face the twin challenges of being scaled up and produced in commercial quantities at a reasonable cost; and still others are speculative, painting a rosy picture of a technology so that the next round of investor funding can be secured.

In many cases, what is not said in an article or white paper is more important than what is said. Developers are reluctant to talk about degradation mechanisms, failure modes, or special maintenance needs, and prospective users may lack the experience needed to ask the difficult questions.

For existing and emerging energy storage technologies, this book provides answers to those difficult questions—and for those technologies not yet announced, a framework is provided to formulate the right questions. Whether the reader is actively considering use of a new storage technology or is simply interested in learning more about these systems, this book is a valuable resource.

Jim McDowall
Saft America
July 2022

Preface

Energy is stored all around us in a host of different forms and for a variety of purposes. Squirrels store acorns for the winter, bears store fat for hibernation, marathon runners store glucose to last 26 miles, and humans store fuel and food for the week or the next seasonal shortage. Today, we store electricity to give us the convenience of using our phones and computers when and where we need them. Just as wires allow electrical energy to move across a distance, electrical energy storage moves it across time and space.

Although the term *electrical energy storage* implies the storage of electricity itself, it can also have a broader definition. Strictly speaking, electrical energy is related to the flow or accumulation of electrons in a magnetic or electric field. In the broader context, electrical energy can be converted to another form of energy and converted back to electrical energy when needed as such.

This book will begin with a short history of the storage of energy using electricity, starting with an electrostatic storage device, and ending with lithium-ion batteries. Chapter 2 describes the main applications of stored energy as they relate to the generation, transmission, and distribution of electricity. Chapters 3, 4, 5, and 6 describe the most common methods to store bulk amounts of energy using electricity (i.e. mechanical, thermal, and electrochemical). Each chapter will include details on the basic science of the storage medium, examples of related products, their pros and cons, and examples of real-world applications of those technologies for commercial purposes. The final chapter will wrap up with a means to compare the different energy storage technologies, types, and products against the value they bring to grid applications.

Humans have been attempting to store electricity for hundreds of years and most recently been very successful at it. There are thousand-page books on the electrochemistry of batteries and many books and countless academic papers on the topic of thermodynamics associated with compressed air. In addition, there are thousands of installations of large energy storage facilities of all kinds, all around the world. The purpose of this book is not to go so deep into every possible area, but give the reader an overall sense of the science behind the technologies described along with some examples, so that they understand their capabilities, limits, and benefits. This book is not intended to answer which is the best technology for today's world-energy storage needs. Rather, it seeks to share tools and experiences to help the reader ask the right questions and find useful answers to what the best technology is for their own applications.

Acknowledgments

I would like to thank the many people who helped contribute to this book in big and smalls ways. A huge thank-you goes to Brianna Hoff, a hard-working chemistry PhD candidate at Princeton University. She spent untold hours editing, revising, and ultimately improving almost every paragraph of my text from the preface to Chapter 7. I am also indebted to Christopher Hoff, a PhD in geology who made invaluable technical contributions to the thermodynamic and electrochemistry chapters. I also appreciate the technical reviews and discussion with Scott Reitsma of ABS, who helped me explain the complex thermodynamic principles in Chapter 5. I want to thank Eli Paster of PolyJoule for giving an early version of the book a once through check for accuracy and literacy and giving me a well-needed boost of confidence. There are many other people who not only contributed photos but reached out to help along the way by reviewing, fact-checking, and catching typos. Finally, I want to share my heart-felt appreciation for my wife, Diane, who without her relentless drive to keep me on my publisher's schedule, this book would not have been possible. She not only kept me on schedule, but she kept me honest, by telling me how much she hated the badly written parts, but also how much she loved the well-written parts. Until the end she kept cheering me on, telling me I could do it, then booting me in the rear to get back to work. I love her to pieces.

CHAPTER 1

A Brief History of Storing Energy Using Electricity

By the mid-18th century, people long knew that there was a mysterious force of nature associated with electrostatic charges. They could generate sparks and miniature versions of lightning by rubbing certain materials together, such as amber ("electron" in Greek) and fur. But once an arc discharged, it was gone. They did not know how to store it for later use. Not that they knew any applications for it other than lighting water on fire, electric kisses [1], levitating light objects, and making the muscles of a dead frog twitch [2].

1.1 Bottled Electricity: First Attempts to Capture Electrical Energy

Toward the end of 1745, a clergyman in Pomerania, Germany, Ewald Georg Von Kleist, attempted to capture electricity in a bottle [3]. One night, he partially filled a glass bottle with water and inserted a metal nail through a cork in the opening and into the liquid. Then, while holding the jar, he touched the nail to his friction machine, which was spun to create an electric charge (Figure 1.1). He thought that perhaps the water would hold the electricity. After doing so, nothing happened: no spark, no fire, no smoke, so he walked the glass jar into another room to get a better look. As he passed through a darkened intervening room, he noticed some corona emanating from the nail in the cork. Curious, he reached out to touch it... WHAM! It hit him. He later wrote to a friend in Berlin:

Figure 1.1 Depiction of the Leyden jar experiment [3].

> I receive[d] a shock which stuns my arm and shoulder. [3]

The jar of electricity became known as the Leyden jar after the town of Leyden, Germany, where a physics professor, Pieter van Musschenbroek and his student Andreas Cunaeus, after each receiving their own life-changing and "horrifying" shocks, eventually figured out that the human body holding the jar was an essential part of the storage mechanism. Improvements on the Leyden jar included wrapping the jar's exterior with a metal foil to serve as the other plate of this newly invented capacitor.

Electricity, whether from lightning, electrostatic discharges, or friction machines, had been a constant source of mystery and curiosity throughout the ages; however, storing electricity was always out of reach until this moment. Just as humans have stored energy in many other forms, such as silos of grain, barrels of pork, firewood sheds, and tanks of oil, people were searching for ways to capture and store this electrical phenomenon and control it. In 1795, Tiberius Cavallo, a physicist and author, presciently wrote:

> In short, nothing contributed to make Electricity the subject of the public attention, and excite a general curiosity, until the capital discovery of the vast accumulation of its power, in what is commonly

called the Leyden phial, which was accidentally made in the memorable year 1745 [3, p. 74].

1.2 Primary Battery Developments

Ever since Von Kleist's accidental discovery, scientists have been attempting to improve energy storage capacity. For many decades, the amount of electricity they could store was not enough to do any real work. From the Leyden jar in 1745 to the Voltaic pile in 1791, stored electricity was used for experimentation, novelties, and sometimes dubious medical therapies [4]. In 1800, Alessandro Volta's Voltaic pile produced a steady stream of electricity [5], but it quickly ran out of energy and was made of materials too expensive to make it commercially viable. It was not until the mid-1800s that the electrical energy stored in a pile of metal layers started providing real value for society. John Frederic Daniell's improvement on the Voltaic pile in 1836 made it practical to power telegraph equipment that sent conversations and data over miles of wires between cities, countries, and continents.

1.3 Early Battery Applications

During the 1800s, engineers and hobbyists were starting to put batteries to work in electric cars and even locomotives. In 1835, a Dutch chemist, Sibrandus Stratingh, created a precarious horseless carriage on three wheels that used a primary battery of copper and zinc plates. A couple of Scots were also busy working on electrified transportation. One of them, Robert Davidson [6], created a couple of electric locomotive prototypes. The second prototype was a 7-ton locomotive driven by two direct-drive motors. At the Royal Society of Arts exhibition in 1841, he demonstrated how it could pull a 6-ton load 1.5 miles. Unfortunately, being nonrechargeable, the batteries had to be replaced at the end of this short run. Not a very stellar performance, but it was a start. Even with such a dismal performance, the local steam-workers union got so upset with the prospect of losing their jobs, they broke into Davidson's shed and destroyed the little engine. It would not be the last time there was open hostility against new technology.

1.4 Secondary Rechargeable Batteries

With the exception of the Leyden jar, until 1859 electrical storage batteries were single-use and not rechargeable. Once these primary batteries were depleted, they were useless as anything else but door stoppers, and had to be recycled to reclaim their constituent elements. They may have been acceptable in the light-duty uses of the telegraph world, but they had limited economic viability in the transportation and power distribution markets. Fast forward to 1854 and Wilhelm Sinsteden invented the first lead-acid battery using two lead plates submerged in a bath of sulfuric acid [7].

It was coined a secondary battery because the electricity that came out of it was the secondary (discharge) current arising from a primary (charge) current being applied to the terminals. Multiple attempts at secondary batteries were made earlier using platinum, tin, lead, zinc, and copper, but Sinsteden's lead-acid combination of metals was the most powerful to date.

Four years later, Gaston Plante improved on Sinsteden's lead-acid battery and made it commercially viable. This version enabled electrical energy to be practically stored at one time and delivered at another. Hence, the start of human's ability to control when and where they use electricity. After 20 years of refinements—the most important of which being in 1880, when Camille Alphonse Faure coated the lead plates with a lead paste—the lead-acid battery became the go-to technology for commercial energy storage.

1.5 Electrified Transportation

Over the next few decades, from 1890 to 1915, batteries sparked a revolution in transportation with electric cars. From 1880 onwards, inventors, hobbyists, carriage manufacturers, and racing enthusiasts created electric vehicles for fun, speed, and business (Figure 1.2).

Rapid advancements in batteries, motors, tires, and chassis were developed in Europe and North America. Electric cars powered by rechargeable batteries dominated the automobile market until 1915, when a combination of poor battery performance, the Great Depression, and Henry Ford's gasoline-powered Model T wreaked havoc on the quiet but clamorous revolution it had been.

Figure 1.2 The first electric production car using rechargeable batteries invented by Thomas Parker in 1884.

The railroad industry was another to put energy storage to use. One of those applications was powering lights in passenger cars when they were stopped in the station or separated from the rest of the train with no other means to power themselves (see Figure 1.3). While the train was moving, the locomotive's wheels would turn a dynamo that powered the lights and charged the batteries. When the train stopped, the batteries would continue to power the lights with their stored energy.

In 1884, Elihu Thomson, one of the founders of General Electric (GE), patented an idea to couple intermittent and inconsistent track-side power with an onboard battery to provide constant traction force for the locomotive (see 1894 U.S. patent #516,666). This invention complemented the growing application of distributed electricity powering trams and trolleys in city streets.

1.6 Storage in Electricity Distribution

In 1880, the first known hydroelectric generating station was built in a chair manufacturing company in Grand Rapids, Michigan. It generated just enough direct current (dc) electricity to light 16 arc lamps for street lighting in the immediate neighborhood. By the end of 1881,

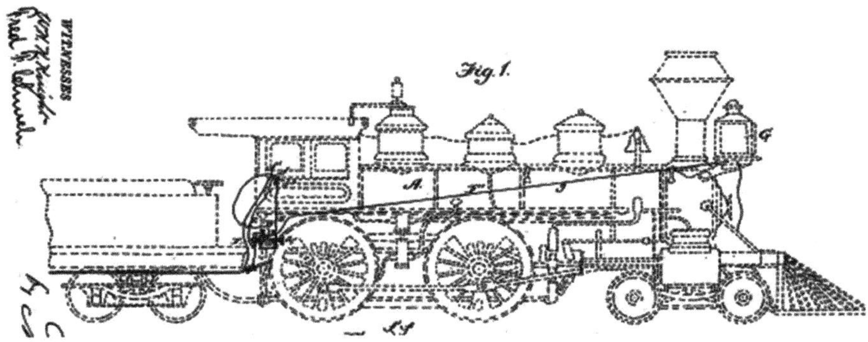

Figure 1.3 Picture from Charles Buell's U.S. patent #259,362. (Source: USPTO.)

streets and factories in cities across North America and Europe were lit up by a lighting system designed by Charles Brush of Ohio. Two years later, in 1882, steam engines powered six dynamos at Thomas Edison's new Pearl Street Station and produced dc current that powered public street lighting in New York City (Figure 1.4). Contrary to public folklore, it was not the first in the world nor the city. Charles Brush's arc-lighting system and steam-powered dynamos powered a stretch of street lighting on Broadway for at least two years before Pearl Street's commissioning.

H. J. Rogers, who owned a paper mill in Appleton, Wisconsin, was inspired by Edison's plans for the coal-burning station and set out to create his own power station powered by flowing water instead of burning coal. In the same year that Pearl Street started operations, Rogers started up the first *commercial* hydroelectric power station that powered his paper factory, his home, and a nearby building. Soon, locally generated electricity was powering many town and city conveniences, such as building lights, trams, trolleys, and even elevators. Although many of the U.S. power plants were distributing dc power, starting in 1885, alternating current (ac) electricity was also beginning to take hold in many cities throughout the United States and especially Europe.

In the effort to sell electrical service, most of the attention was paid to distributing it geographically throughout the cities, suburbs, and rural residences of the population. However, the problem that added cost and waste to the means of producing and distributing electrical power was that consumer loads were not constant. For example, homes used electricity to power lights mostly at night, paper mills and machine shops

Figure 1.4 A Pearl Street station model as shown in the March 2, 1929 issue of Electrical World, p. 459.

mostly operated in the day, and fans only ran when it was warm. Consequently, power plants, wires, generators, transformers, and any other supporting infrastructures were sized to handle the maximum probable load and therefore were much larger than required for a constant average load.

Edison's power plant charged its very first New York customers by the number of light bulbs they had in their house, regardless of whether they were ever turned on [8]. Today, many utilities charge an extra fee from large commercial and industrial customers for drawing excessively above their average current in peak periods of the day. These demand charges could be more expensive than the actual energy charge depending on the utility and the customer's peak demand profile, but are used to compensate the utility for the excess infrastructure they require to deliver peak load.

To give some perspective on how much variability exists on a typical electrical grid, Figure 1.5 shows an example of the seasonal variability

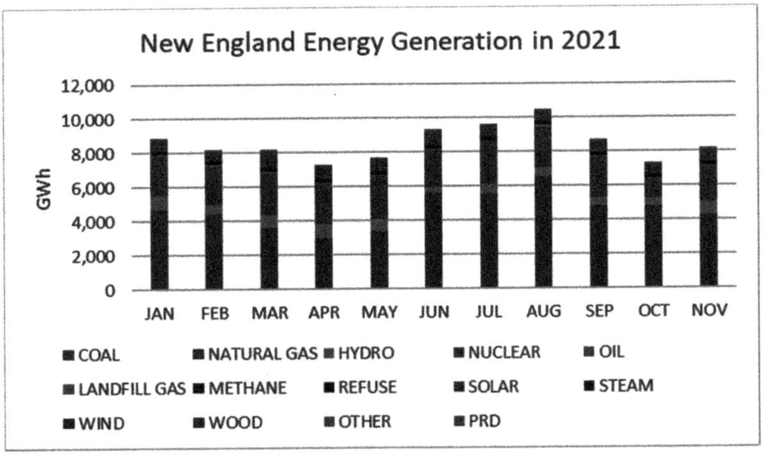

Figure 1.5 Chart of monthly energy generation and usage in New England [9].

of electricity usage in an eastern region of the United States in 2020. During the heat of July, New England residents used over 10,000 GWh of energy in one month, while in the shoulder season of April, when it was not too hot or cold, they used just 7,000 GWh.

The hourly consumption rate is equally variable and cyclic during a typical daily profile. Figure 1.6 shows daily power fluctuations where the daily electrical power demand in New England ranged between 12 GW and 20 GW [10].

Since the early 1900s, behind the scenes and scattered around the world, inventors have been thinking of how to distribute electricity over time so that generators can run at their maximum efficiency and wires can be utilized to their optimal levels. Even Edison recognized that there could potentially be improved efficiency by employing a battery at the point of loads to level the consumption over a daily cycle. In an interview with *The Electrician* magazine in 1883, just one year after he opened Pearl Street Power Station in New York [11], he said:

> In 1879 I took up that question, and devised a system of placing storage batteries in houses connected to mains and charging them in the daytime, to be discharged in the evening and night to run incandescent lamps. I had the thing patented in 1879 (I forget the date of the patent), but there is nothing in it. (See U.S. patent #218,167 [11].)

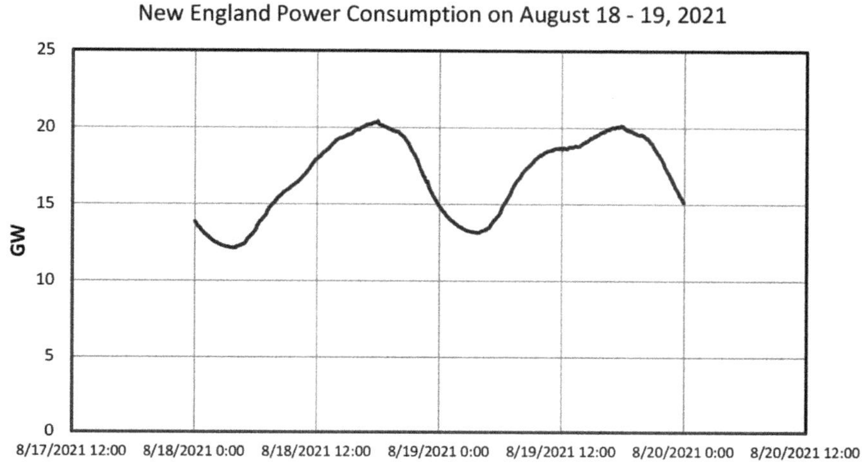

Figure 1.6 Daily load fluctuations in New England [10].

Later in the interview, Edison went on to say how batteries of that day were not up to the challenge of daily cycling and would not be commercially viable:

> Scientifically, storage is all right, but, commercially, as absolute a failure as one can imagine. You can store it and hold it, but it is gradually lost, and will all go in time. Its efficiency, after a certain number of charges have been sustained, begins to diminish, and its capacity and efficiency both diminish after a certain time in use, necessitating an increased number of batteries to maintain a constant output [11].

He was unabashedly harsh on batteries and their purveyors. This was the same interview in which Edison was quoted saying:

> Just as soon as a man gets working on the secondary battery it brings out his latent capacity for lying [11].

That sentiment did not stop him and others from thinking of and patenting ideas for how batteries could stabilize intermittent power loads and generation. In 1885, Schuyler S. Wheeler invented an elevator that used the energy stored in a battery to lift a cart and then used the cart's motion downward to recharge the battery. This had a positive effect of reducing the peak power drawn from the utility and improving the entire system's efficiency (see patent #US324621A). Between 1896 and 1910, William Lord Bliss authored dozens of patents related to improving the electric supply in utility and applications (see patents #572,627, 601,233 (Figure 1.7), and 799,516 as examples).

Charles Brush, a prolific inventor and businessman, is best known for his advancements of the dc dynamo and arc-lighting systems of his day. At its peak, Brush's company supplied half of all the lighting systems to the major cities in Europe, North America, and even Japan. In 1888, he constructed a wind-powered generator in his backyard (Figure 1.8). The 60-ft tall structure powered a 12-kW dc dynamo that was electrically coupled to batteries in his basement. His was the only powered home in Cleveland at that time, and powered it was! He had 350 incandescent lamps, 2 arc laps, and 3 motors. When the wind was going strong, it would charge the batteries. The batteries would, in turn, power the loads even when the wind was not blowing. This may be the first example of batteries being applied to integrate renewable energy into a grid.

1.7 Pumped Hydro Energy Storage

Across the ocean, Europeans were working with a different kind of energy storage—one that could be commercialized and last much longer than batteries. In the industrial sector of Zurich, Switzerland, engineers built one of the first known pumped hydro storage facilities designed to

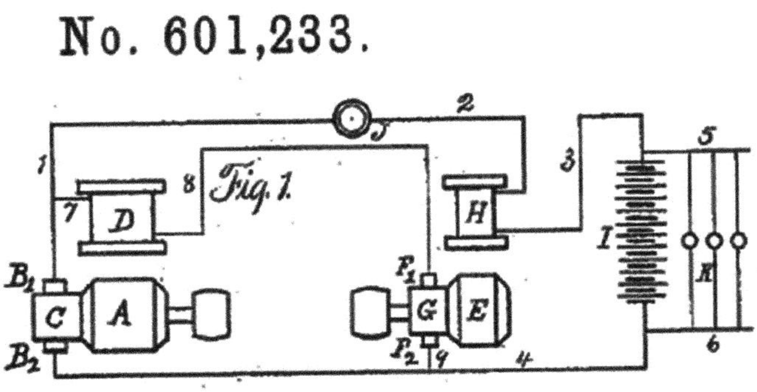

Figure 1.7 Picture from Lord Bliss' patent #601,233. (Source: USPTO.)

Figure 1.8 Windmill in Charles Brush's back yard, Cleveland, OH, 1883.

move bulk amounts of energy across time. Prior to and at the time of its construction in 1882, waterpower from the Limmat River was used to mechanically power a thriving industrial sector in the city (Figure 1.9). Over 1 km of cables extended from the water turbines to the industrial centers to power their machines through an extensive arrangement of gears and clutches.

However, the river's water ebbed and flowed with the seasons, and sometimes the machines slowed to a crawl, stifling productivity. The city's waterworks authority then conceived of an idea to use excess mechanical energy in times of abundance to mechanically pump the water up into a reservoir. In times of drought, the Oberwasser was channeled back down through chutes to boost the mechanical energy provided to the city.

Less than 10 years later, in 1891, it was converted to an electromechanical facility using motors and generators to harvest and store the water's mechanical energy.

Figure 1.9 Zurich Waterworks: turbines and pumphouse, 1882. (Limmat-Clübler information brochure, 2016.)

1.8 Growth of Hydro-Based Storage and Generation

For the next 40 years, Europeans built 40 more pumped hydro facilities before the United States installed its first significant energy storage facility in New Milford, Connecticut [12]. The Rocky River Pumped Storage Hydraulic Plant was built between 1927 and 1929 along the Housatonic River in Connecticut (Figure 1.10). At that time, there was about 30 MW of electrical power capacity at several hydrogeneration facilities along the Housatonic, but it was not reliable. Like the Limmat River, the water flow varied from season to season and even year to year, depending on rainfall and weather patterns. Engineers thought of ways to regulate the flow and fill gaps in generating capacity. They finally settled on a design that pumped water from the river up to a man-made lake 200 ft above the river. The pumps and motors were reversible, so they could use a single penstock to transport the water up or down between the two locations.

Such a project was not without controversy since it had to displace homes, schools, churches, cemeteries, and wildlife, all of which had to be uprooted and relocated, including the interred. This was in addition to 4,500 acres of trees and brush that had to be cleared by hand as well.

Elsewhere in the United States, hydroelectric generation facilities, such as the Bonneville, Grand Coulee, Niagara Falls, Hoover Dam, and

Figure 1.10 Rocky River pumped hydro station. (Source: FirstLight. Reprinted, with permission, from [13].)

several in the Tennessee Valley Authority region, among others, were built to generate electricity. Over 2,000 hydroelectric generation plants were built in the United States between 1910 and 2014. Although this amounted to almost 80 GW of power, it was still not enough capacity to power the growing U.S. supereconomy.

Coal, nuclear, gas, and oil-powered plants were built to meet the rest of the demand. In 1950, about half of all electrical power in the United States was generated by coal-fired power plants and remained about that percentage until the 2000s when it started to fall out of favor. Currently, it contributes less than 20% of the total electricity consumed by the United States (see Figure 1.11). In the 1960s, nuclear power started to add its might into the mix. It saw a steady rise until the 1990s when it contributed about 20% of the total energy into the U.S. power market and continues to do so today.

Around 1950, given the rise of coal and nuclear, and understanding the need to keep those plants running at a steady clip, utilities had a renewed interest in energy storage. Starting in the mid-1950s, utilities started building pumped hydro storage facilities wherever possible. There are now about 153 facilities in the United States [15], ranging in

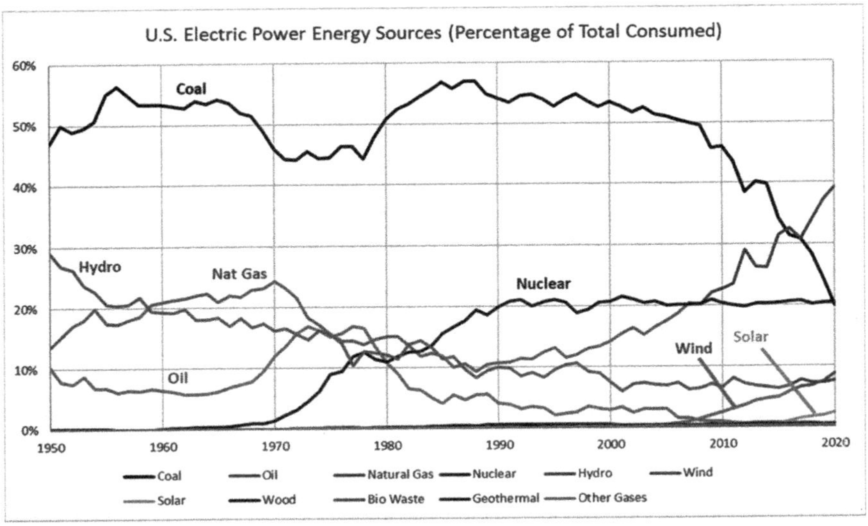

Figure 1.11 U.S. electrical power contributions from 1960 to 2020 [14].

size from 8 MW to 3 GW, dedicated to cycling energy on a regular basis. The total amount of pumped hydro storage in the United States is now 22 GW and comprises about 95% of the total energy storage in the country.

1.9 Compressed-Air Energy Storage

Until 1978, all substantial grid-service energy storage in the world was based on pumped hydro. Then, in Germany, Brown, Boveri, and Cie did something entirely new. They built a 290-MW compressed air energy storage (CAES) facility (Figure 1.12) that took excess energy from a nearby nuclear plant and stored it in the form of compressed air in man-made underground caverns [16]. The caverns were 250m tall into which they could pump air up to 70 bar. The facility takes 8 to 12 hours to charge at a rate of 60 MW and, if augmented with natural gas, can output 290 MW for 2 to 3 hours into the electrical grid. It is still in operation today, and the only other one like it in the world is a 110-MW CAES facility built in Alabama in 1991.

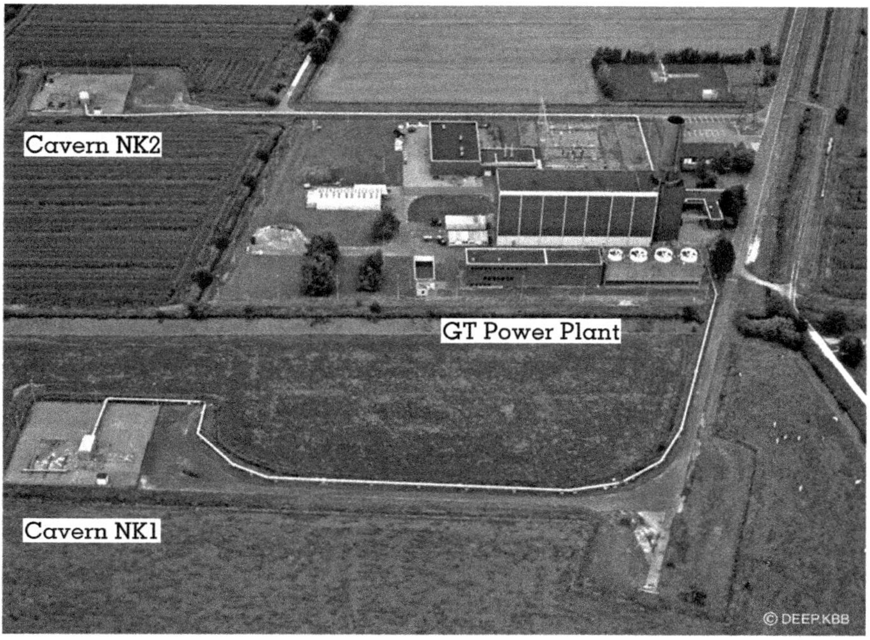

Figure 1.12 Compressed air energy storage facility in Germany. (© DEEP. KBB GmbH.)

1.10 Return of Battery-Based Energy Storage

In the 1980s, after about 20 GW of pumped hydro energy storage was installed in the United States, the Electric Power Research Institute (EPRI) and the U.S. Department of Energy (DOE) started to study alternative energy storage means. One of their first test trials was a 500-kW/kWh lead-acid battery in California performing peak-shaving service. Peak-shaving is the process of providing power to a load or the grid during times of heavy use, so that power through one or more grid components can be reduced for an economic or technical benefit.

The DOE's lead-acid storage system was built and evaluated in California in 1983, then transferred to a working site in North Carolina. There, it clipped the peaks off the power imported into a small municipal utility to help offset their demand charges. This modestly sized energy storage system saved the small utility over $200,000 in 7 years. Ironically, the death of this battery was not caused by overuse, but an errant charging circuit that completely damaged the battery cells in 1999, giving credence to the saying, "Old batteries never die; they are murdered."

The second attempt at lead-acid grid energy storage was in Chino, California, at the Southern California Edison utility in 1988 [17]. A joint venture between EPRI and DOE funded the installation of a 10-MW/40-MWh lead-acid battery (Figure 1.13). Its intent was to evaluate how well a battery could perform in starting up other generators after an outage, trimming the peak loads off critical infrastructure, and assisting in the stabilization of grid voltages. It operated as expected for the first couple of years, performing more than 150 discharge cycles, each of which provided more than its rated energy. However, the facility was used only intermittently for a few years after that, before being decommissioned in 1999.

Since then, according to the DOE Energy Storage Database [19], 52 more lead-acid storage facilities were installed in the United States with a combined power of 143 MW. Of those, 43 remain in operation, but with only a total capacity of 52 MW.

Figure 1.13 Inside the Chino lead-acid 10-MW/40-MWh energy storage facility [18].

As more renewable energy sources, such as wind and solar, were introduced into the electric grids in the 2000s, interest in energy storage was again renewed. The output power from renewable energy sources is neither predictable nor continuous. Therefore, energy storage is required to enable them to supply power when it is needed by the grid (see Chapter 2). As a result of this new interest, many other energy storage technologies have been tested and used by utilities, independent power producers, and enterprising private organizations. Batteries, thermal storage, compressed air, flywheels, and even more pumped hydro storage were installed all around the United States and the world. Numerous types of batteries were developed, trialed, and put to substantial use, including lead-acid, sodium-sulfur, lithium-ion, nickel-metal hydride, zinc, vanadium redox, and other chemistries. In total, the DOE Energy Storage database records about 26 GW of energy storage installed in the United States between the first pumped storage facility in Milford, CT in 1929, to the latest lithium-ion project in California (Figure 1.14).

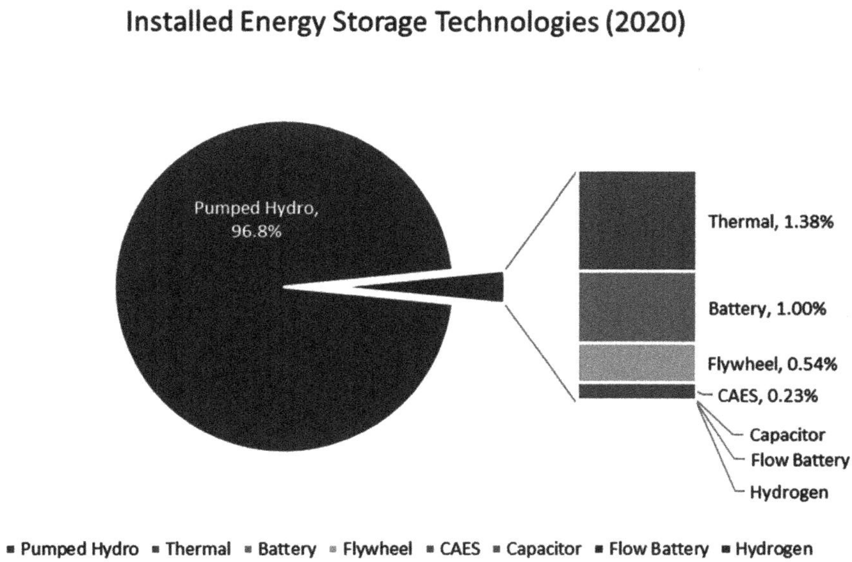

Figure 1.14 Graph of energy storage technologies in use around the world in 2020. (Data from DOE GESDB [19].)

1.10.1 Lithium-Ion Batteries

Since 1992, the development of the lithium-ion battery revolutionized batteries for the portable devices market. Its theoretical and practical capacity to store energy far exceeded that of any incumbent technology. In the early 2000s, laptops, mobile phones, and personal digital assistants (PDAs) had mostly switched to lithium-ion from nickel-cadmium (Ni-Cd) and nickel-metal hydride batteries (NiMH). In the effort to create a product that had the most run time, battery manufacturers were focused on how much energy they could fit into the smallest package. High power capability was not at all necessary since a battery with 12 hours of run time has an inherently low power draw.

In another corner of the consumer marketplace, power tool makers were eyeing lithium-ion technology to improve the lackluster run times, anemic power, and short service life that Ni-Cd batteries then offered. Since the only lithium-ion options were cells designed for laptops having an 8- to 12-hour run time, they were hampered by the low power capabilities of these cells. They looked for a cell that could give them more power without having to spend so much effort cooling them.

In 2005, Yet-Ming Chiang, Bart Riley, and Ric Fulup started a company called A123 Systems in Watertown, MA with several ideas on how to make a better lithium-ion battery. One of their ideas, inspired by the work of John Goodenough at the University of Texas, was a way to improve the phosphate molecule that offered a lower-cost alternative to the more common lithium-ion ingredient of cobalt oxide. What resulted was a lithium iron phosphate (LFP) cell that had extremely high discharge rates and a very long cycle-life. Whether it was the chemistry, the particle size, the conductor designs, or the cell design, their cell could discharge in a few minutes, not overheat, and do it all over again thousands of times.

Using this cell, Black and Decker created an industrial version of their DeWalt tool line (Figure 1.15), a 36V battery pack, that made their tools even more powerful than some of their own corded versions, let alone the rest of the industry's cordless products.

Chris Shelton, of The AES Corporation, was also intrigued by the DeWalt battery packs, not for the tools but because of the A123 Systems 26650 power cells in them. Shelton's company was a large independent power producer (IPP) that provided grid regulation services using

Figure: 1.15 Black & Decker's 36V industrial power tool using A123 Systems Nanophosphate cells.

gas-fired combustion turbines that could start up and deliver instant power at a moment's notice. In those days, an instant was about 10 to 15 minutes, the time it took to start a turbine, connect to the grid, and deliver power. If they wanted such a resource to react quicker, they would have to leave it spinning and connected to the grid, wasting a lot of idling energy.

Shelton thought high-power batteries, like A123's, could start instantly and not waste energy during idle times. It could also get its energy from the grid when power was abundant and even help to stabilize the grid in the process. The challenging aspect of this and other grid applications is that the power source must cycle many times during its service life. Ordinary batteries performing this service would soon wear out and need frequent replacement. A123's cells were unlike the typical lithium-ion batteries of that time in that they could provide high power and many cycles.

In 2007, Shelton introduced his idea to A123 and, 9 months after a contract was signed, A123 tested the world's second[1] grid-sized

1. The first was Altairnano's lithium-titanate 1-MW, 250-kWh container also commissioned by AES, but in July 2008.

lithium-ion energy storage system at the AES Huntington Beach Power Plant in California, in November of 2008. It contained over 83,000 cells, like the ones in the DeWalt power pack, capable of providing 2 MW of energy for over 15 minutes of run time and performed just as Shelton dreamed it would.

Since then, AES, A123, Saft, Fluence, Tesla, Nidec, BYD, Doosan, and a host of other enterprising battery manufacturers, developers, utilities, and independent power producers have built energy storage systems around the world using similarly capable lithium-ion cells (Figure 1.16) to perform grid stabilization services, peak shifting, demand management, and many other grid services.

The list of energy storage projects continues to grow. Installations of energy storage have rapidly accelerated since 2010. When the first batteries were installed, 10 MW was considered a large battery. In fact, the 10-MW, 40-MWh SCE battery in Chino was the largest battery-based energy storage facility at the time. Then the 20-MW in Puerto Rico took the title in 1994, followed in 2003 by the 27-MW Ni-Cd battery in Fairbanks, Alaska. By 2020, battery-based energy storage systems were regularly designed to be 100s of MW in size, and a GWh worth of energy on a single site was not unheard of.

In 2020, LS Power, traditionally a solar-power generation developer, installed the then world's largest 250 MW/MWh battery energy storage system in southern California. As of this writing, they were actively pursuing plans to install 500 more MWh of energy to give it 3 hours of

Figure 1.16 Various energy storage sites installed around the world from 2009 to 2021. (Source: NEC. Reprinted with permission.)

energy at 250 MW. Within a year of this feat, Vistra, an independent energy producer, started operating a 300-MW, 1200-MWh battery-powered energy storage facility in a refurbished gas-fired combined-cycle power plant in northern California (Figure 1.17). At the time of this writing, this is the largest battery energy storage system built. Other projects planned around the world will eventually take the title.

1.11 Going Forward

In the past few decades, growing evidence of the correlation between global warming and CO_2 emissions and other by-products of the world's energy production has led the drive to reduce the use of fossil fuels to satisfy our energy needs. Policy at the city, state, and national level in many countries have led to serious efforts to map out a carbon-free roadmap in the next few decades. Because energy storage enables widespread adoption of renewable energy, it is being developed and installed at a record pace all over the United States and the world. In the following chapters, we will review many of the energy storage technologies that have been developed, installed, or will be available in upcoming years.

Figure 1.17 Inside the world's largest energy storage facility in 2021: Vistra's 300-MW / 1200-MWh BESS. (Source: Vistra Corp. Reprinted with permission.)

References

[1] Bertucci, P., "Sparks in the Dark: The Attraction of Electricity in the Eighteenth Century," *Endeavour,* Vol. 31, No. 3, 2007, pp. 88–93.

[2] Lyons, M., "Galvani Discovers 'Animal Electricity'," *History Today,* Vol. 71, No. 9, 2021.

[3] Houston, E. J., *Electricity in Every-day Life,* Vol. 1, Philadelphia, PA: Collier and Sons., 1904, pp. 71–73.

[4] Houston, E. J., *Electricity in Every-day Life,* Vol. 3, Philadelphia, PA: Collier and Sons, 1904, pp. 548–575.

[5] Decker, F., "Volta and the 'Pile'," January 2005, https://web.archive.org/web/20120716205546/http://electrochem.cwru.edu/encycl/art-v01-volta.htm.

[6] Reid, J. S., "Robert Davidson – Pioneer Electrician," *The Scientific Tourist: Aberdeen.*

[7] Lockerby, P., "Wilhelm Sinsteden–Inventor of the Lead-Acid Battery," *The Chatter Box,* May 16, 2013.

[8] Arnold, T., "A Short History of Demand Charges (And How to Trim Yours)," October 4, 2011, https://gridium.com/history-of-demand-charges/.

[9] New England ISO, "Net Energy and Peak Load," 2021, https://www.iso-ne.com/isoexpress/web/reports/operations/-/tree/net-ener-peak-load.

[10] New England ISO, https://www.iso-ne.com/isoexpress/.

[11] *THE ELECTRICIAN,* Vol. iX, 1883.

[12] ASCE, "Rocky River Pumped Storage Hydraulic Plant," 2020, https://www.asce.org/about-civil-engineering/history-and-heritage/historic-landmarks/rocky-river-pumped-storage-hydraulic-plant.

[13] FirstLight, "Rocky River Pumped Hydro Storage Station," 2022, https://www.firstlightpower.com/facilities/?location_id=354.

[14] EIA.gov, "Electricity Net Generation, 1950 - 2020," 2020, https://www.eia.gov/totalenergy/data/monthly/pdf/sec7_4.pdf.

[15] EIA.gov, "Existing Capacity by Energy Source, 2019," 2019, https://www.eia.gov/electricity/annual/html/epa_04_03.html.

[16] Crotogino, F., "Huntorf CAES: More than 20 Years of Successful Operation," Spring 2001 Meeting of the Solution Mining Research Institute, Hannover, Germany, 2001.

[17] EPRI, "Chino Battery Energy Storage Power Plant: First Year of Operation," *EPRI,* 1993.

[18] Rodriguez, G. D., W.C. Spindler, and D. S. Carr, "Operating the World's Largest Lead/Acid Battery Energy Storage System," *Journal of Power Sources,* 1990, p. 10.

[19] Sandia National Laboratory, "DOE Global Energy Storage Database," 2020, https://www.sandia.gov/ess-ssl/global-energy-storage-database-home/.

Additional References and Useful Links

http://www.americaslibrary.gov/jb/gilded/jb_gilded_hydro_1.html

https://en.wikipedia.org/wiki/Charles_Grafton_Page

https://www.powermag.com/history-of-power-the-evolution-of-the-electric-generation-industry/

https://www2.census.gov/prod2/statcomp/documents/HistoricalStatisticsoftheUnitedStates1789-1945.pdf

https://www.milrose.com/insights/an-early-history-of-lighting-in-nyc

https://web.archive.org/web/20090224192642/http://chem.ch.huji.ac.il/history/brush.html

http://edisontechcenter.org/index.htm

https://www.edison.com/content/dam/eix/documents/innovation/smart-grids/Energy-Storage-Concept-to-Reality-Edison.pdf

https://www.eia.gov/electricity/annual/html/epa_04_03.html

https://www.eia.gov/totalenergy/data/monthly

CHAPTER 2

Applications of Energy Storage on the Grid

Chapter 1 presented the origins of energy storage, from the Leyden jar to the electrochemical storage cell. This chapter details the applications of energy storage on the power grid and answers the following questions:

- Why does the grid need energy storage?
- How much energy storage does the grid need?

2.1 Why Does the Grid Need Energy Storage?

The grid is a system of interconnected generators, loads, wires, power conversion devices, and energy storage that together serve to create and deliver electricity to consumers. A grid can be as small as a privately owned microgrid consisting of solar panels, batteries, a generator, and household appliances. Larger grids, serving entire countries or even continents, connect thousands of generators of many types with loads ranging from street lighting to homes, factories, and city transportation systems. In all cases, the generators are designed and coordinated to provide a consistent, expected voltage and frequency,[1] so that the connected loads can safely and reliably pull power from it.

1. In this book, a grid is an ac system where the voltage and currents travelling through the network oscillate at a specified frequency (50, 60, 440, etc.). Grids can also be dc systems where power is distributed by nonoscillating currents at different voltage levels set by distributed electronic power converters. (Read more about ac versus dc in Appendix 2A.)

Each grid performs a balancing act between power delivered and power received. In a grid without any storage, power going in must exactly match power going out. Control systems generally keep this balance in check, but if sudden surges in available power or grid load occur, the grid's voltage and frequency will at best see unacceptable levels or, at worse, collapse to zero. Without a means to temporarily store power surges from either supply or demand, the resulting variations of the grid's voltage and frequency could cause unacceptable power conditions for its connected loads.

Grid operators balance the delicate supply and demand through a variety of means, including predictive modeling, resource scheduling, financial incentives for electricity consumers and generator operators, ancillary service markets, and day-by-day management of their connected resources. Energy storage has always been one of the many tools that operators use to manage temporary differences in supply and demand, from the instantaneous response of inertial energy storage to the slower responding, bulk energy reserves. As the grid changes from being primarily powered by large, centralized bulk power stations to distributed, intermittent, inertialess renewable resources, energy storage is becoming increasingly vital to the grid's health. Energy storage in many forms is used for a variety of grid services, such as inertia support, frequency regulation, frequency response, renewables integration, ramp management, energy shifting, transmission and distribution support, power quality, resiliency, and more. The following sections describe each of these services and provide some examples to illustrate their benefits.

2.1.1 Inertial Support

Traditional, grid-connected generators spin in unison and are synchronous to the ac frequency. Prime mover torque exerted on the generators' axles will push electrical power into the grid and incrementally affect the frequency. Likewise, power pushed into or out of a generator causes it to push or pull against the force of the prime mover, affecting its rotating speed. The synchronized rotation of many spinning masses is a key element of the traditional grid.

In addition, each of the spinning generators on a grid has built-in energy storage due to the inertia of the prime mover turbines, flywheels, and generator rotors. This inertial energy is useful in keeping the grid stable. Small perturbations in the electrical load and generation will not significantly upset the rotational speed and the operating frequency because of this inertia.

The energy stored in a spinning mass is:

$$E_{rot} = \frac{1}{2} J\omega^2 \tag{2.1}$$

where J is the moment of inertia of the spinning mass, and ω is the rotational speed in radians per second.

The inertial constant of a machine is defined as:

$$H = \frac{E_{rot}}{S_{rated}} \text{ (seconds)} \tag{2.2}$$

where S_{rated} is the power rating of the machine.

The differentiation of (2.1) in combination with (2.2) results in a convenient way to estimate the changes in frequency when a large difference between generated and consumed power, ΔP is suddenly experienced in the grid

$$\frac{df}{f_0} = \frac{\Delta P}{S_{rated}} \frac{dt}{2H_{sys}} \tag{2.3}$$

where H_{sys} is the system's rotational inertia, the sum of all the individual connected spinning rotational inertias, while S_{rated} is the power rating of all the connected generators.

The swing equation (2.3) shows that the greater the imbalance in power, the greater the shift in frequency, while larger inertia and system size will decrease the frequency change for a given ΔP.

Typically, a large generator synchronized to the grid has about two to five seconds worth of energy stored in its spinning mass when operating at its rated output power. The amount of inertial energy of a particular generator is dependent on its spinning mass as well as its physical design. The amount of inertial energy on a grid at any one moment depends on how many and what types of machines are connected to it. Because the number and type of generators changes over the course of a day, week, and even season, the grid's inertia varies significantly, as well. For example, Finland's grid load ranges between 6 to 14 GW, as shown in Figure 2.1, and the estimated inertia ranges between 100 to 250 GW-sec.

This is around 28 to 70 MWh of energy stored in the spinning mass of its generating stations and any synchronized spinning motors connected to the grid. The chart in Figure 2.1 shows that the grid's inertia generally follows the load trend during the year. In general, higher loads correlate to more connected generators, which will increase the inertia.

Most grid operators simulate the operation of their grids to determine the minimum acceptable inertia required to maintain their operating performance with respect to frequency fluctuations. They factor in typical load conditions and common scenarios where major power plants may suddenly stop contributing to the grid's power supply. These scenarios are called contingencies and can be caused by any number of unexpected failure conditions inside and outside of the power plants.

A large grid, like one that spans a continent and links several countries, will have tens of thousands of large spinning generators, all synchronized to the same frequency and providing inertia into the grid. Larger grids tend to have more stability because the relative size of a single contingency is smaller. For example, the three grids in the United States are listed in Table 2.1. The smallest grid, Texas, has the largest

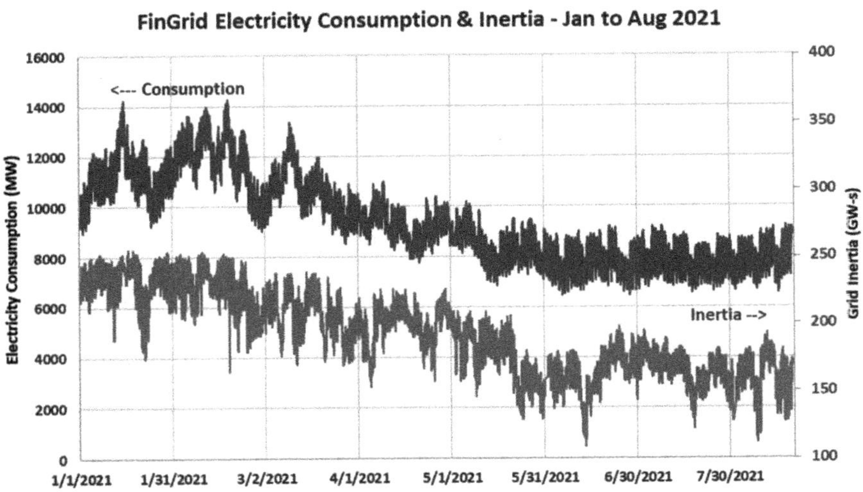

Figure 2.1 Electrical energy consumption and grid inertia in Finland in 2021. (Date source: www.fingrid.fi.)

Table 2.1 Contingency Size of Three U.S. Grids

Grid Name	Peak/Avg. Demand (GW) (2018)	Largest Contingency (MW)	Largest Contingency (% of Average Load)
Western	168/100	2,626	2.6
Eastern	556/354	4,500	1.3
Texas	74/43	2,750	6.4

potential contingency of 6.4%, while the largest grid has a potential contingency of 1.3%.

2.1.1.1 Inertial Stability

Figure 2.2 represents what happens when a contingency occurs on a grid. Frequency starts at 60 Hz at $t = 0$ when the contingency occurs. It falls at a rate calculated by (2.3) and, if no frequency response assets respond to the event, the frequency will continue along the "Base Inertia" curve to intersect the under-frequency limit setting (UFLS) threshold in 2.5 seconds. If adequately sized primary frequency response (PFR) as-

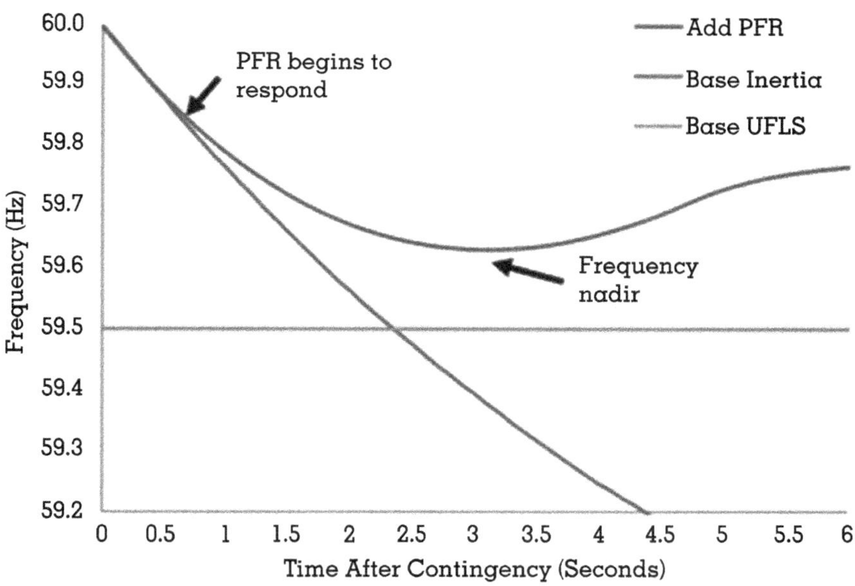

Figure 2.2 Representative time-chart of the system frequency after a contingency, with and without primary frequency responses. (From: [1].)

sets react quickly, the frequency change will decrease to a nadir (lowest point), and then eventually stabilize at some point below the nominal 60-Hz level (following the "Add PFR" curve).

The UFLS (and an equivalent over-frequency limit (OFLS)) is a low-frequency threshold at which relays will open breakers in tie-lines and generating stations. Therefore, grid operators do everything they can to keep the frequency from crossing the UFLS and OFLS thresholds. System inertia and size, fast-acting PFR, and setting the thresholds all effect the system's stability.

2.1.1.2 Grid of the Future Is Now

Since large-scale rotating generators have built-in inertia they can provide short-term stability for large grids. However, inertialess power generators, such as photovoltaic (PV) and many newer wind generators, electronically inject power into the grid, synchronized to the grid's frequency. When the grid's frequency decreases due to a contingency, the electronic power convertors just follow the declining frequency and continue to deliver the power they were commanded before the event. Because these inertialess generators do not try to influence the frequency in any way, they have little stabilizing effect on it. Grids such as those in California, Texas, Germany, and Denmark have a high percentage of inertialess power generation and require other means to stabilize their frequency during contingencies. The following are measures that could be taken to ensure frequency stability:

1. Include more renewable energy generation that uses synchronous generators such as hydroelectric, geothermal, solar-thermal, and others.
2. Include energy storage technologies that use synchronous generators such as pumped hydro and compressed air.
3. Include a mix of distributed synchronous condensers, which can also serve to regulate voltage and affect power flow when necessary.
4. Employ inverter-based generators for fast frequency response, reserving a portion of their power capacity for up and down regulation.
5. Deploy fast frequency responsive battery-based energy storage.

6. Control several non-time-critical loads to switch on and off in response to frequency deviations.
7. Deploy partially loaded, grid-forming, inverter-based generators coupled with energy storage.

Regarding #7, as renewable generation penetrates more than 50% of the grid's load, a whole new approach to grid control will be needed. Instead of having inverter-based generators follow the grid's frequency, it will be necessary for them to form their own internal reference and set their output voltage and frequency target using GPS synchronized time and/or network communications. When multiple generators are interacting on a single grid, one challenge is to keep them coordinated and not fight each other in achieving their own reference frequency and voltage. The other challenge is when sudden load increases occur on a fully commissioned grid. Electronic inverters that are maxed out will not be able to provide an extra boost of power to help stabilize the voltage or frequency. To prepare for these contingencies, it will be necessary to either reduce the continuous output of the grid-forming generators or deploy fast-acting energy storage to provide up and down regulation. This type of inertialess, grid-forming control has been operating in many smaller microgrids for decades [2], so it is possible, but needs to be further explored before it is deployed on a massive, grid-wide scale.

2.1.2 Frequency Services

The grid's inertia is useful for common occurrences of short, minor perturbations of its generation and loads; however, when large power plants suddenly drop out of service, or tie lines between major portions of the grid unexpectedly open, the grid can experience very large frequency disturbances. With no mitigation, these disturbances can ripple throughout the grid and cause large area power outages. Grid operators usually have allocated a portion of their generators' capacity to respond in seconds to such disturbances.

2.1.2.1 Frequency as a Health Indicator of an AC Grid

In any ac grid, its frequency shows the instantaneous difference between supply and demand. According to the swing equation (2.3), when the frequency is unchanging, the supply and demand are balanced. When the frequency is decreasing, there is a deficit in supply; when it is increasing, there is more supply than demand.

2.1.2.2 Frequency Regulation

Assets that help support the grid frequency are either frequency regulation or frequency response systems (FR for both). There is a subtle difference, but they essentially perform a very similar service. Frequency regulation systems respond to a signal from the regional utility or utility collective (such as an independent system operator (ISO)) that, according to the needs of that specific portion of the overall grid, tells the asset how much power to apply to or absorb from the grid at any given moment. The entity controlling the signal reads the grid frequency, applies an algorithm, and then sends out a signal to all the frequency regulation assets under its jurisdiction [3]. This signal is called the automatic generation control (AGC) signal. The advantage is that the controlling entity can be more selective about which asset it uses and when it demands this service for each event. By monitoring the whole grid at once, they have a broader perspective about the grid's health at different locations.

The grid frequency has always fluctuated due to millions of loads turning on and off at the whim of consumers, but as more inertialess and unpredictable renewable energy sources come online, the frequency will vary even more. Traditionally, the more flexible fuel-fired generators in a utility's fleet have responded to the frequency regulation signal, some of them performing better than others. Recently, fast-reacting batteries and flywheels have been deployed to perform this service. Their performance has been so accurate that the grid can use less of them (with respect to power) than they did when using power plants to achieve the same level of frequency control.

Because the frequency fluctuates so much, the AGC signal causes the participating energy storage systems to send and receive power in and out of the grid many times a day. To illustrate an example of this duty profile, Figure 2.3 shows the AGC of a representative regional ISO for part of a typical day in 2019. The demanded frequency regulation power on this day in January 2019 rapidly fluctuates up and down, in a seemingly chaotic manner. The midpoint of the vertical axis is zero power. Above the midpoint, the FR asset is commanded to absorb power from

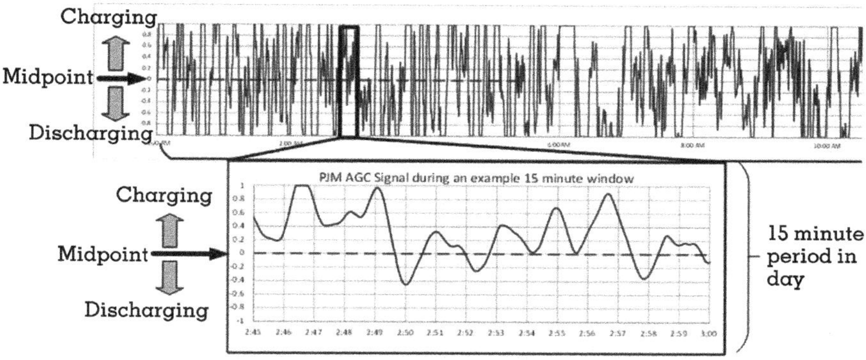

Figure 2.3 PJM AGC Signal on a typical day in 2019. (Data source: PJM.)

the grid (charge), and below the midpoint the FR asset is commanded to deliver power to the grid (discharge).

The chart shows that in both charging and discharging modes, the ISO commanded the FR assets to continuously vary their power contribution between 0% and 100%. A zoomed-in, 15-minute window is shown to clarify how the signal varies minute by minute. In this 15-minute window, the control signal switches between charge and discharge six times. Slower responding and fixed-power energy storage systems will struggle to follow this signal and maintain a high performance score.

Performance score is a metric that numerically ranks how well FR assets follow an AGC signal. A regional utility or ISO will send an AGC signal to all the FR assets, and the FR assets are expected to control their output power levels to this signal. Every ISO or utility managing FR asset is different, but in the Pennsylvania Jersey Maryland (PJM) ISO's case, the formula for how PJM judges the performance and therefore the compensation of its FR assets is:

$$Score_{Perf}(t) = \max_{i=0 \text{ to } 5\text{min}} \left[A \cdot Score_{delay}(t+i) + B \cdot Score_{corr}(t+i) \right] + C \cdot Score_{precision}(t) \tag{2.4}$$

More details can be found in the PJM handbook for frequency regulation [4]. The mathematics behind this formula are complex but, in summary, if the shape of the FR asset's response curve is correlated to that of the AGC signal, its delay following the AGC is minimal (seconds), and if its 10-second delayed magnitude is the same as that of the AGC, the performance score will be very high.

There are many examples of flywheel and battery-based energy storage used for frequency regulation in the deregulated U.S. markets. In fact, 300 MW of regulation has been deployed in PJM in the United States, 80% of which are batteries, and the rest are mostly flywheels [5]. Examples of these are presented in Chapters 3 and 6.

2.1.2.3 Frequency Response

The disadvantage of the centrally controlled frequency regulation process is that there may be a delay in detecting a frequency event, the processing of the signal, the transmission of the signal, and the response time of the asset. The whole process can take seconds at best, and sometimes can be affected by unreliable communications and connections.

Frequency response assets monitor the local grid frequency and can respond within milliseconds according to a prearranged response curve. A typical algorithm programmed into an energy storage system is shown in Figure 2.4.

In some utilities, power plants have performed this service by operating at a lower output power than their maximum power levels to leave

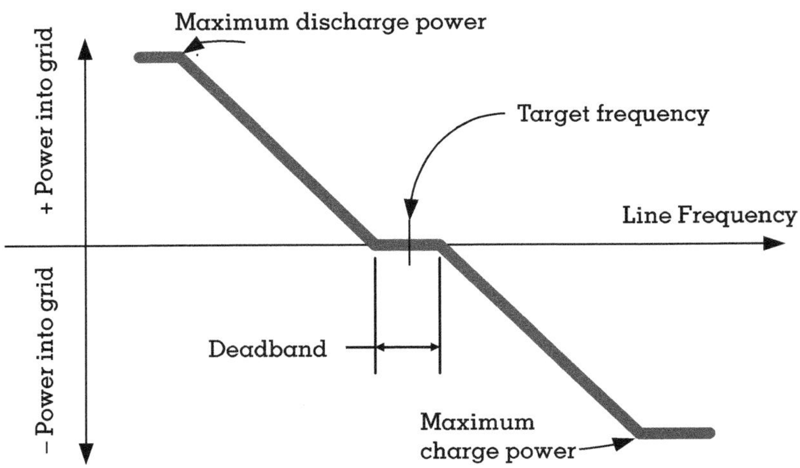

Figure 2.4 Frequency response asset output power vs grid frequency.

room at the top for reserve power. These assets are already spinning in synchronization with the grid and, when needed, they burn more fuel and push power into the grid. These assets are called primary or spinning reserves because they are spinning along with the grid and can react within seconds to a contingency to provide the primary response to grid disturbances. They can also react to sudden excess power conditions by decreasing their power output.

Other assets can be held on standby, offline, and not spinning. When called into action, they will need to start their engines, spin up, get synchronized with the grid, connect to it, then begin pushing power into it. This can take several minutes, so they are not as responsive as the spinning reserves, but they save fuel by not running idle continuously. These are called secondary and tertiary reserves, and they can only deliver power to the grid, not absorb it.

The chart in Figure 2.5 shows a scenario where a large power plant was suddenly taken offline, and primary, secondary, and tertiary assets were called into play to help stabilize the frequency in the seconds and minutes following the event.

One example of a frequency response energy storage application is in Chile. In 2009, AES installed a battery-based frequency response energy storage system made by then A123 Systems (Figure 2.6). Power plants in the region, mostly coal at that time, reserved 5% of their power for frequency response services that had several associated problems:

1. Based on the utility regulations in that region, they made no revenue from the frequency reserve services they performed. Their profits were lower because of the unsold capacity.
2. They needed to keep the boilers hot enough to run at full capacity, but they were not outputting the full capacity except when the reserves were requested. So, they were running less efficiently and spewing more carbon into the atmosphere per GWh than if they continuously ran the plant at full capacity.
3. They were not very good at it. Large power plants change their power in tens of seconds while the frequency changes every second.

By offloading the frequency regulation services to a battery system, the power plants could run at full capacity, provide baseload power to

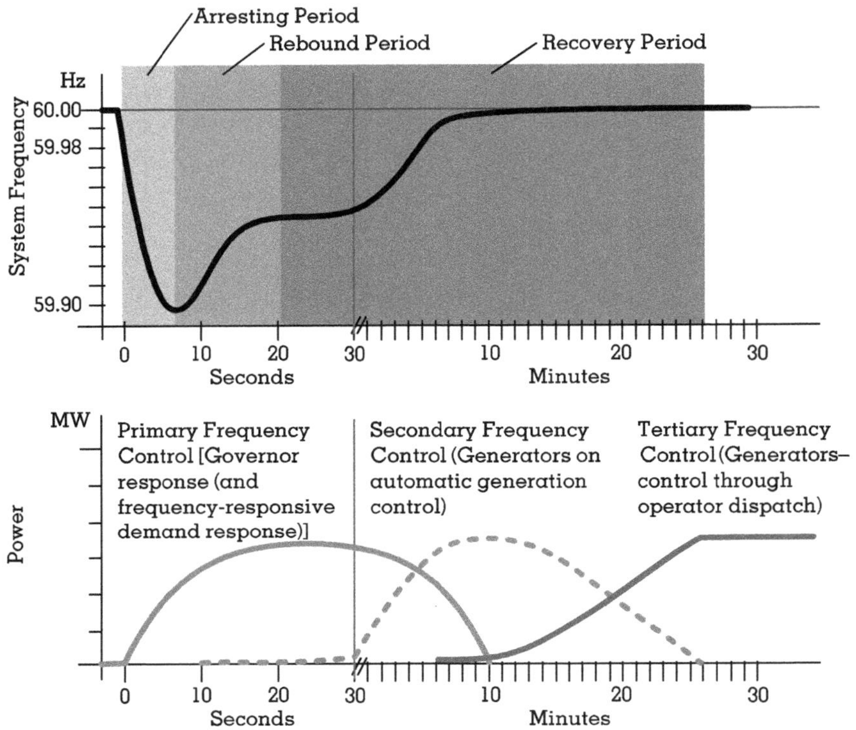

Figure 2.5 Frequency response assets in action. (© 2010 The Regents of the University of California, through the Lawrence Berkeley National Laboratory.)

the region, make more profit, and operate more efficiently. In addition, since the batteries respond in milliseconds to frequency events, they are much more effective at stabilizing the frequency and grid's operations than the large power plants.

2.1.3 Renewable Energy Integration

Renewable energy power sources, such as PV and wind, not only fail to contribute to the grid's inertial energy reserves, but they also contribute to an increased amount of grid power variability. The sun is often blocked by clouds, while wind regularly shifts up and down and rarely blows constant. Figure 2.7 shows a snapshot of the power output from a medium-sized PV array in upper New York State.

Figure 2.6 12-MW frequency response asset located in the Atacama Desert in Chile. (Source: NEC. Reprinted with permission.)

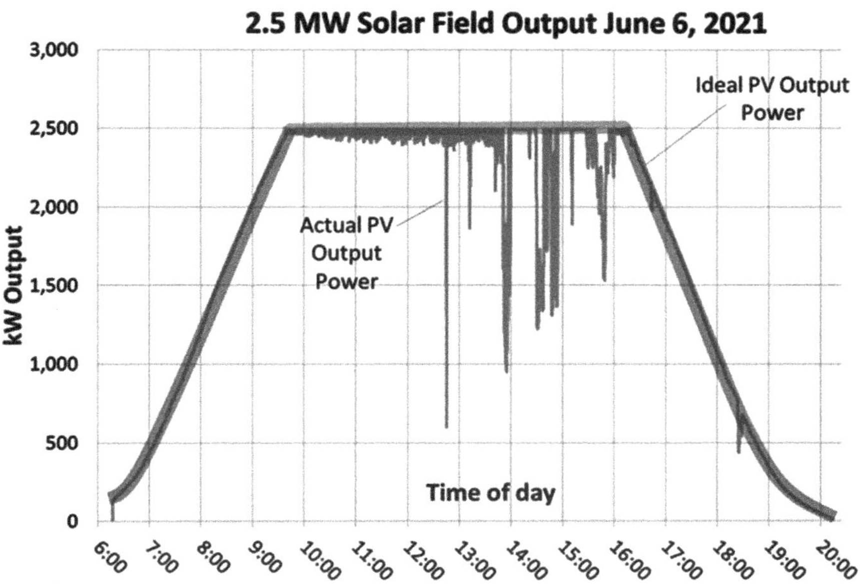

Figure 2.7 Medium-sized PV array power output on June 6, 2021.

This is a good day! The perturbations of generation between noon and 4 PM on this day are deep but very short. During other days, large sections of the daily output are not available because of deep cloud cover, such as the example of a couple of days later, shown in Figure 2.8.

Needless to say, the power injected into the grid by these resources is unreliable as it is subject to nature's whims. When PV and wind resources are only a small part of the grid's generating capacity, their fluctuations can be absorbed by the natural inertial reserves of the larger grid's spinning, generating assets. However, as the number of renewable generators increases, the grid becomes more and more dependent on fast-reacting energy storage reserves to stabilize it in everyday operations.

The Sterling Municipal Light Department (SMLD) was faced with its own problems associated with having too many renewable energy resources. In 2013, it had 3.2 MW of PV generation installed and operating. With only 3,200 customers, this represented about 30% of their peak load. If the PV generation looked anything like the charts in Figures 2.7 and 2.8, the power coming in from the regional grid would be equally chaotic to compensate for the variable PV power.

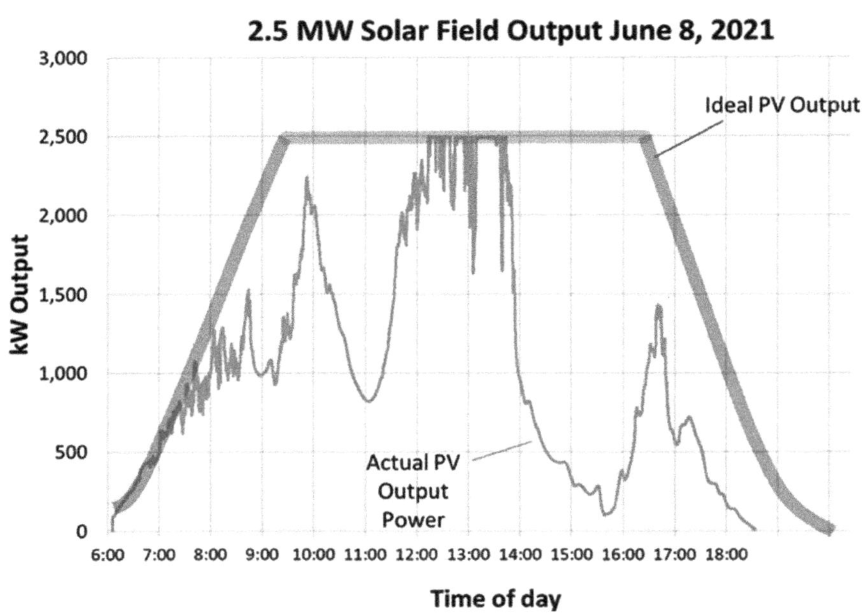

Figure 2.8 Medium-sized solar array power output on June 8, 2021.

Consider a realistic example in which SMLD's customer load is 10 MW on a hot and steamy day in July. The PV array is churning out 3 MW, and the regional grid is providing the difference of 7 MW. Suddenly, a cloud covers the region, and the PV array drops its power output to 1 MW, a third of what it was. At that point, the regional grid needs to suddenly deliver another 2 MW on top of the 7 MW it is already supplying. On this example day, the regional grid may also be running at full capacity, so the extra 2 MW may cause unexpected stress on generators, transmission lines, or other resources. This stress may or may not be technically managed by the equipment, but it will certainly cause SMLD to have to pay extra fees toward its electric bill. Because this scenario played out multiple times after SMLD installed their 3.2-MW PV array, SMLD was hit with millions of dollars of additional demand fees from their regional grid. Sean Hamilton put a moratorium on additional PV resources in the SMLD territory until they found a solution to this problem.

To help reduce their peak power demands from the regional grid, SMLD installed a 2-MW/3.9-MWh system from NEC Energy Solutions for $2.5 million (Figure 2.9). After its installation in 2016, they not only reduced their demand-related fees, but they were able to operate their grid with more renewable energy, saving their customers twofold.

In another example, the island of Maui, Hawaii was in the process of installing 22 MW of wind power. The only problem was that this amounted to 10% of the island's peak load. When the wind farm generates electricity, its power goes up and down, almost as chaotically as PV arrays do, as shown in the charts in Figures 2.7 and 2.8. Worse yet, when hurricane–force winds blow on the turbines, they sometimes need to shut down instantly to avoid damage. This chaotic power generation causes the island's diesel generators to have to react more quickly than they were designed to do. If those generators react too slowly to a sudden change in wind generation, the power quality on the island suffers. The frequency and voltage will vary outside of their required limits and may even trigger outages when breakers overreact to these unusual variations.

Energy storage was installed to smooth the power variations from the wind farm in a process called ramp management, shown in Figures 2.10 and 2.11. As the power from the wind farm rapidly changes in magnitude, energy storage delivers or absorbs energy to reduce the actual

Figure 2.9 2-MW/3.9-MWh battery energy storage system in Sterling, MA. (Source: NEC. Reprinted with permission.)

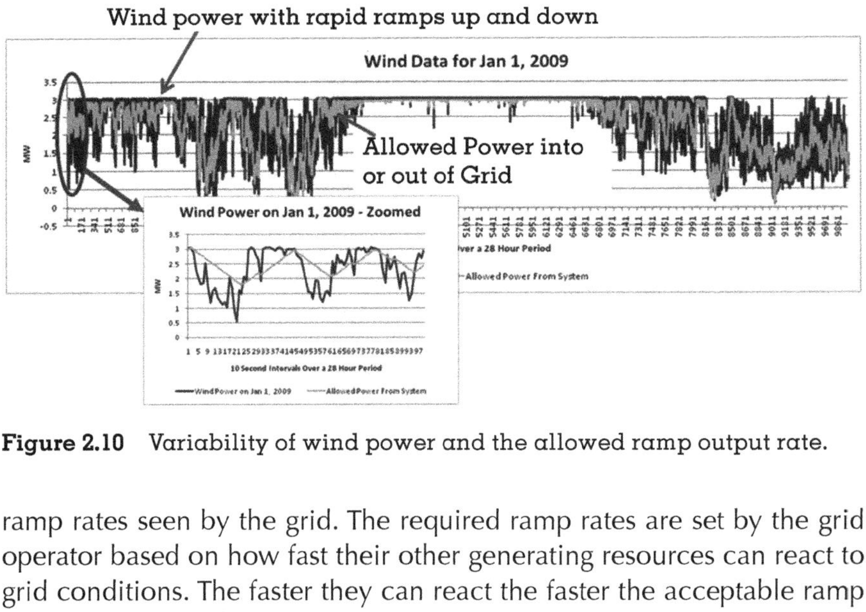

Figure 2.10 Variability of wind power and the allowed ramp output rate.

ramp rates seen by the grid. The required ramp rates are set by the grid operator based on how fast their other generating resources can react to grid conditions. The faster they can react the faster the acceptable ramp rates can be.

2.1.4 Energy Shifting

Technically, every application using energy storage is, by definition, energy shifting. However, one is particularly well-known and recognizable for its consistent behavior. The California duck curve in Figure 2.12 is a plot of the net electricity demand, the total load minus wind and solar energy for a few challenging spring days in the years 2018 to 2021.

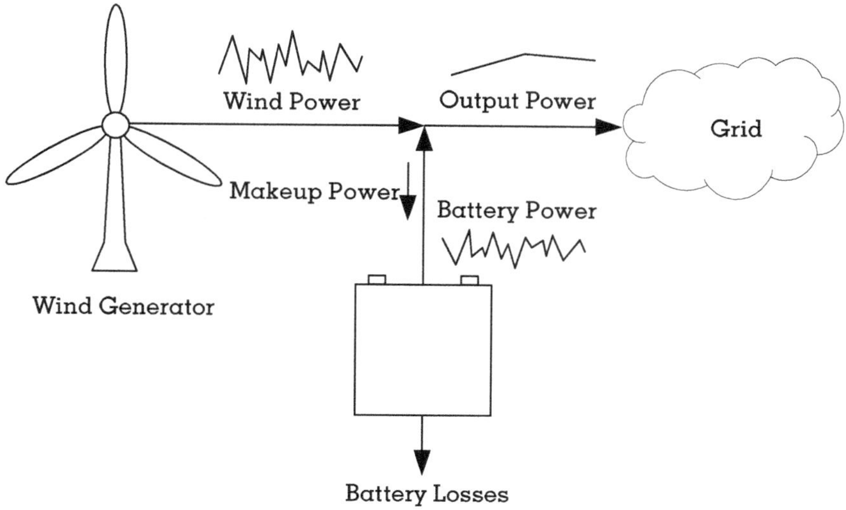

Figure 2.11 Diagram showing how energy storage interacts with a wind generator to smooth its output into the grid.

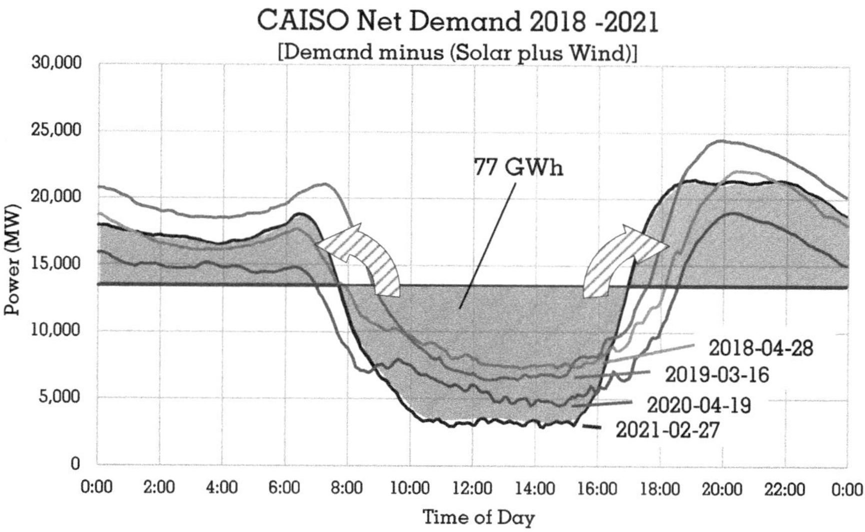

Figure 2.12 California ISO duck curve, 2018–2021. (Data source: CAISO. http://www.caiso.com/TodaysOutlook/Pages/index.html.)

The data shows that the traditional energy resources plus imported power from neighboring utilities have very little demand during the sunny hours of the day (around 3.5 GW in 2021), then quickly shift to a very high demand (21 GW) during the late afternoon to early evening hours, and finally settle down to a moderate level (17 GW) at night. This type of duty profile on a traditional fuel-fired power plant is very stressful, having to go from very low to extremely high in a matter of hours, and repeat every day. The chart also shows that the trough is getting deeper every year, and could someday go negative, causing a tremendous amount of curtailed renewable energy.

Finally, one can see the sheer magnitude of the energy involved in just one region's grid issues. In order to move all of the peak energy to the trough to achieve perfectly level power from nonrenewable resources, the energy storage would have to move 77 GWh on this particular day (Feb 27, 2021). Given the current cost of energy storage, a number of approaches to "tame the duck" are likely to be considered, including time-of-use incentives and multiple types of energy storage. However, even if energy storage was free, its round-trip losses of the energy transferred on a daily basis would amount to multiple GWh of wasted energy. The cost of energy storage and its operating losses must be weighed against the benefits of smoothing the curve, and it is likely that a mellowing of the curve may be the best compromise, not a full flattening of it.

2.1.5 Transmission and Distribution

Variability of customer electricity usage is inevitable in daily profiles such as in the hour-by-hour profile of a typical day in Figure 1.6. The grid infrastructure must be sized for the peak demand so that it can support its quality-of-service commitments to the utility's customers. That means that all the utility's assets required for supporting the peak load are oversized with respect to the average customer loads. If energy storage can somehow be deployed to level the power being delivered by the utility's infrastructure, the infrastructure could be reduced in size and cost. Certain niche cases, such as those in vacation hotspots, will likely see even more dramatic changes from season to season throughout the year.

In some situations, the cost of adding infrastructure to deal with temporary peak loads is larger than the cost of deploying an energy storage system that could reduce infrastructure strain. The peak-rated infrastructure could be as simple as an additional transmission line, or it could also be an upgraded transformer, a strategically positioned generator, or a larger power plant.

In one example, in 2018, the island of Nantucket off the coast of Massachusetts was predicting an ever-increasing peak summer demand that would soon outstrip its undersea power cable's capacity [6]. The option of adding another undersea cable over 30 miles was compared to that of adding a battery—the battery was cheaper. The estimated cost of an additional cable was $200M [7]. National Grid, the island's utility company, installed a 6-MW, 48-MWh battery energy storage system (BESS) (Figure 2.13) to handle the peak summertime loads that occur when 40,000 visitors descend on the small island of 11,000 people. The cost of the battery at that time, including installation, was a little over $80M, so the utility saved $120M in up-front costs.

As battery prices have been coming down, especially from 2010 to 2020, there are increasingly more examples in which a battery-based energy storage system is less expensive than adding new infrastructure to support a growing peak demand.

2.1.6 Demand Charge Management

Utilities have an incentive to deploy energy storage to help stabilize their grid and reduce peak loading, and many customers do too. Many utilities charge customers for drawing a large amount of power for brief periods of time.

To avoid these demand charges, customers can deploy energy storage behind their meter and program it to deliver power when their

Figure 2.13 48-MWh battery energy storage system in Nantucket, MA. (Photo courtesy: National Grid, United States.)

demand is high and slowly recharge when their demand is light. This delivered power offsets the power being used by the customer, such that the peak demand on the utility is reduced to an acceptable level. With certain types of load profiles, and in certain utility markets, the savings can be dramatic and actually pay for the energy storage system in a few years.

A demand charge is a general term, used on a customer level, for any fee associated with peak power usage. At the utility scale, these charges are referred to as network service charges and forward capacity market payments. Whatever the name, the concept is similar. Customers and local utilities are charged a fee for the ability and/or the past use of power, in addition to their past use of energy.

One example of a customer reducing its demand charges using energy storage is the SMLD installation in Massachusetts. The customer in this case is a municipal utility in the small town of Sterling that buys power from the larger regional grid at bulk energy rates. SMLD then sells to its local customers' energy at a reduced rate. The rates at which it buys electricity are low, but if it takes too much power from the regional grid at one time, it can be charged a demand fee that they will continue to have to pay for the rest of the month, even if they do not continue to demand that amount of power. It just takes one hour in one day where the power exceeds a certain threshold to force them to pay a fee for the rest of the month. To help reduce their peak power demands from the regional grid, SMLD installed a 2-MW/3.9-MWh system from NEC Energy Solutions for $2.5 million.

On one particularly cold day, just before the 2-MW energy storage system was commissioned, Sean Hamilton, the general manager of SMLD, noticed that their estimated import power was going to hit a record high that day and likely induce an expensive demand fee. He called NEC and asked for a special favor. He asked if they could discharge the battery at the expected peak time of day. Chris Quaranta from NEC took the call and, after making a few internal calls to the engineers, decided to help his customer. Figure 2.14 shows the power draw by SMLD from the regional grid during this event.

The peak at around 6 PM was absorbed by the NEC battery system for about two hours. As a result, they saved SMLD's customers $17,000 in just two hours. The following summer, on June 13th, SMLD would

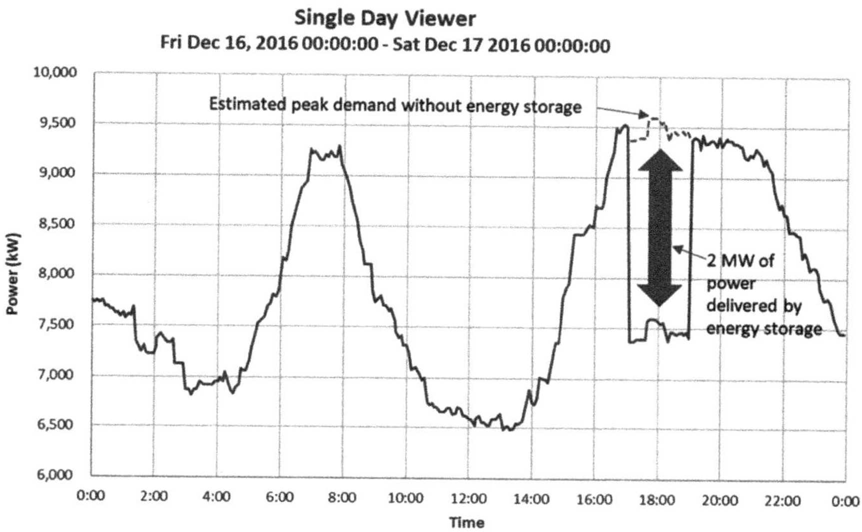

Figure 2.14 Peak power and energy reduction by SMLD during a winter peak.

have experienced another surge in demand from the regional grid, this time from air-conditioning loads. Again, they used the NEC battery to absorb the load during the two-hour period around when they expected the peak (Figure 2.15), and this time it saved the ratepayers over $240,000.

In fact, the regular use of that battery for this reason has saved the town of Sterling ratepayers an average of $400,000 a year.

2.1.7 Power Quality

Power quality is another application of energy storage that covers the assurance that voltage and frequency are maintained within certain bounds for acceptable operation by the majority of loads plugged into the grid. It also includes other aspects that could be deleterious to loads or even annoying to people, such as harmonics, dc content, flicker, and step-changes in voltage [8].

In addition to providing frequency support services as previously discussed, energy storage can support local voltage quality by injecting either real or reactive power into the grid to correct voltage conditions near its connection point. Real power comes from the energy storage, so

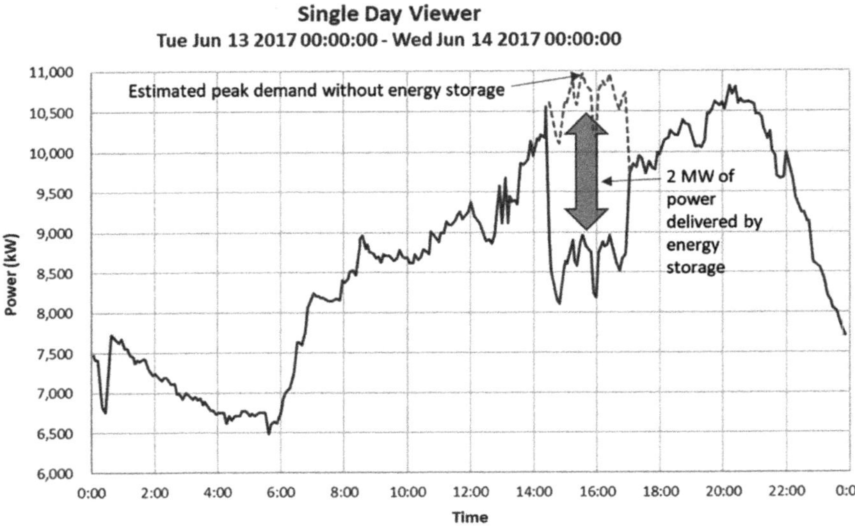

Figure 2.15 Peak power and energy reduction by SMLD during summertime peak.

it can only be used for momentary conditions. Reactive power takes no net energy from the energy storage, so it can be used continuously up until the total reactive power (VAR) limitations of the inverter. Similar to frequency response systems, energy storage can be programmed to respond to voltage conditions in a precoordinated pattern such as in Figure 2.4, but instead, the x-axis would be voltage and the vertical axis would be reactive power.

One good example of an energy storage system improving power quality is one that was installed in Alaska for the Golden Valley Electric Authority (GVEA). In December of 2018, a massive earthquake hit the coast of Alaska, knocking out a 170-mile-long powerline that supplied a substantial portion of GVEA's power needs from the Anchorage area utility. Instantly, GVEA's 40-MW BESS kicked in and provided the 30 MW of power that was previously supplied by the intertie. The BESS is only rated to supply a few minutes' worth of energy at that power level, so GVEA operators immediately started up a 60-MW backup generator to supply the region's power until the powerline could be repaired. GVEA manager Pete Sarauer said it is likely that few if any of its customers were even aware that the co-op was scrambling to avoid a blackout [9]. The Ni-Cd battery system was installed in 2003 by Saft and ABB and was intended to provide local power quality support for the area [10]. The long intertie from the Anchorage utility was prone to damage from weather and, because of its length, led to voltage disturbances when the region's loads changed abruptly. Having the local battery provide instant voltage support, spinning reserves, AGC response, and instantaneous power scheduling has paid off for this power co-op. Since 2003, it has responded to an average of 50 major grid disturbances each year [11] and dramatically improved its customers' power quality.

2.1.8 Resiliency

One definition of resiliency is the ability to respond constructively to unexpected challenges. Energy storage can provide this benefit to power grids with advanced preplanning, coordination, and the right management. As discussed earlier in this chapter, energy storage can be very useful in stabilizing the grid. It can deliver and absorb power into the grid when it detects a need to do so. Its responsive manner has a stabilizing effect on the grid's frequency, the local grid voltage, and the power flow in the region. Energy storage can also be used to bolster portions of the grid for brief periods of time after a loss of a generator or transmission line, while backup means to provide the electricity are called into service.

Additionally, energy storage can be configured to provide emergency power to localized service areas and critical load during times when there is no power available from the grid (i.e., blackouts). Naturally, individual energy storage sites cannot hold up the entire grid but, if preconfigured properly, energy storage can back up critical loads and smaller portions of the grids, such as neighborhoods, apartment buildings, and even single-family houses.

When a grid outage occurs, a small piece of the grid (microgrid) can be isolated by opening a switch (Figure 2.16), and the energy storage can provide power to the microgrid for as long as the power outage lasts, or as long as its energy reserves hold out.

2.1.8.1 Black Start

Another way that energy storage can provide resiliency is in black start services. This is when an energy source can power a portion of the grid

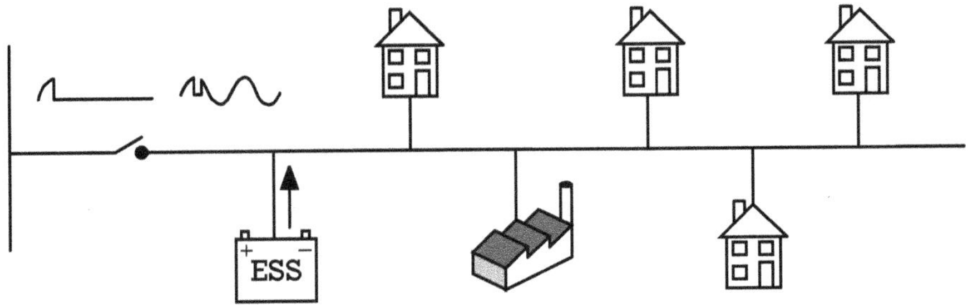

Figure 2.16 Simplified representation of an islanded microgrid after a grid power disturbance.

in a grid-forming mode so that other power generators can start up. Some generators need auxiliary power to start spinning and get in sync with the grid before they can start accepting power from their prime mover. Energy storage capable of providing grid-forming power can perform this service, in addition to its other primary functions, when called to do so.

2.2 How Much Energy Storage is Necessary?

Looking at the fundamental drivers for energy storage in the United States, 25% of the carbon released into the air comes from electric utilities [12]. Reducing the country's carbon footprint will require a significant reduction of emissions from this sector. Replacing coal-fired plants with gas-fired plants is one step in the right direction but, ultimately, grids need more power from renewable energy sources.

If the grids are predominantly powered by renewable resources, there will be a need for much more energy storage than currently installed. Considering the California ISO example, a hot day in August of 2021 shows a peak demand of 42 GW in that region. On that day, 15 GW of renewable energy was being deployed, including solar, wind, and others, but not including hydro. The remaining power came from gas, coal, nuclear, hydro, and imports (from neighboring utilities). If the region were to eliminate coal, gas, and imports, it would have to expand its renewable resources by four times, accounting for its diurnal behavior. Since this new fleet of energy supply is not coordinated with the demand, about 26 GW of energy storage would be required to move 230

GWh of energy from the middle of the day to the evening hours (Figure 2.17).

This assumes a round-trip efficiency (RTE) of 85%: batteries will be higher, pumped hydro, CAES, flow, and others will be lower. As of 2021, this region deploys around 1.5 GW of energy storage to move around 5.4 GWh of energy on peak demand days. Based on this one day, to allow this region to go (almost) 100% renewable, it will have to deploy about 40 times the energy storage it currently has.

This hypothetical exercise on historical data from August 2021 is representative of only a fraction of the U.S. power grid, which is a fraction of the world's energy usage, yet its numbers are eye-opening by themselves.

National Renewable Energy Laboratory (NREL) published a series of reports [13] that discuss a broad range of energy storage topics from technology to markets to costs and values. One of the reports estimates the total U.S. energy storage requirement in 2050 to be between 1.3 TWh and 6 TWh, depending on which economic scenario is assumed [14]. The U.S. load in August of 2021 peaked at around 700 GW.

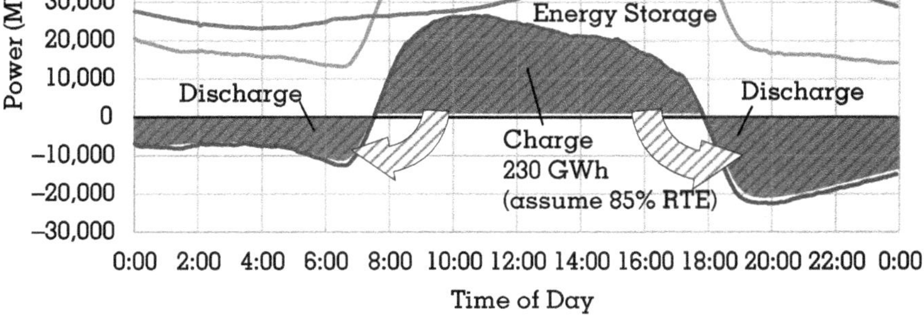

Figure 2.17 CAISO region supply and demand scenario where renewables, nuclear and hydro, power the region's loads. The difference to be made up between supply and demand on this day is 230 GWh (including round-trip losses.) (Historical data source: CAISO.)

According to NREL, the peak power in 2050 will be twice that of 2021, or about 1.4 TW. The ratio of NREL's estimate of 6 TWh and 1.4 TW, is 4.3. This roughly aligns with the CAISO analysis above which says that the amount of energy storage will be 5.5 times the peak demand. Using these two examples, one can approximate the amount of storage needed to completely decarbonize the grid with an assortment of wind, solar, hydro, nuclear, and others, as a ratio of 4 to 5.5 times that of the peak summertime load.

To understand the full extent of global energy storage requirements, similar studies should be applied to every region accounting for resource availabilities and load profiles. Other regions will have different mixes of renewables, for example, Texas, Germany, and Denmark have more wind than solar power, Iceland has geothermal energy, and Canada has hydroelectric energy. Load profiles will vary as well. Florida and Arizona have extraordinary daytime air-conditioning loads while the Nordic countries and Russia have high nighttime winter loads. Other areas like Detroit and New England have a wide variability between summer and winter seasons.

Furthermore, there will be extenuating circumstances where long periods of resource inadequacy are experienced. If clouds cover an entire region for several days, PV generators' outputs will be crippled, and the region's power will need to be imported from neighboring regions if they even have any power to spare. Energy storage will have to be oversized to bridge long gaps in renewable generation that may occur from day to day and even month to month. Energy storage products that can provide multiple weeks of energy will be required for these situations, while more efficient, shorter term energy storage will be used for daily cycling.

Between 2019 and 2020, for the first time ever, the cost of installing a PV generating plant became lower than that of a conventional gas-fired plant [15]. The caveat is that this does not include energy storage. Adding storage to the PV mix makes it still more expensive than conventional power generation as of 2020. However, going forward, as the prices of energy storage fall, renewable energy becomes more viable. As renewable energy becomes more widely adopted, more energy storage is required. As more energy storage is manufactured and sold, their prices come down (Figure 2.18), which supports their increased viability and adoption rates.

NREL published another report describing five phases of energy storage deployment in the United States [17]. Although focusing on the United States in particular, this pattern is representative of many developed countries around the globe. Phase 0 was prior to 2011, where massive amounts of pumped hydro storage was installed between 1900 and 2011, first in Europe, then the United States and the rest of the world. Twenty-three GW of storage were installed in the United States in this phase. Phase 1 was the deployment of high power, low energy, fast-reacting energy storage for frequency regulation, and other grid stabilization services. Although only a few GW of storage were installed in this phase, there is still a need for roughly 25 GW of these services in the United States Some of that need will be serviced by longer-duration storage in the next phases. In phase 2, because of declining prices, batteries capable of 2 to 6 hours start to replace peaker plants and serve in the energy capacity markets. By the time this phase ends, NREL expects between 30 and 100 GW of storage installed in the United States. Phase 3 is where longer duration, 4- to 12-hour energy storage will be provided by even lower-cost batteries and other technologies such as flow batteries, modernized compressed air energy storage (CAES), and efficient pumped hydro storage plants. NREL expects more than 100 GW

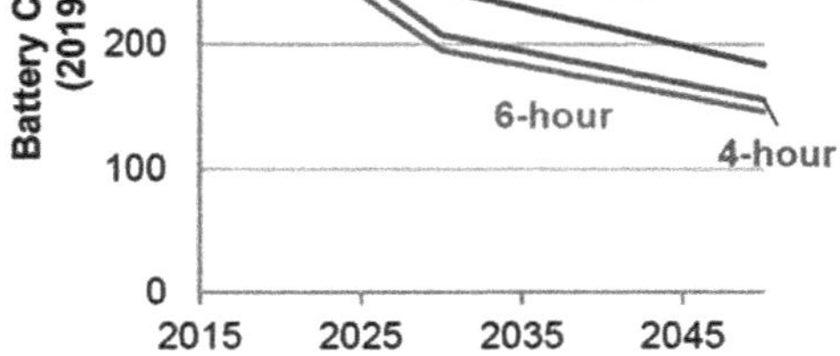

Figure 2.18 Falling prices of lithium-ion battery installed cost. (From: [16].)

of storage, suited for daily energy shifting needs of mostly PV generation, will be installed in this phase. The final phase is what NREL dubs, "The End Game—Multiday to Seasonal Storage." Storage of this phase is used to carry excess energy from one season of the year to the next and will include over 250 GW of ultra-low cost, not necessarily efficient, gravitational, electrochemical, chemical, and thermal energy storage.

References

[1] Denholm, P., et al., "Inertia and the Power Grid: A Guide Without the Spin," May 2020, https://www.nrel.gov/docs/fy20osti/73856.pdf.

[2] Ackermann, T., et al., "Paving the Way: A Future Without Inertia Is Closer Than You Think," *IEEE Power and Energy Magazine*, Vol. 15, No. 6, 2017, pp. 61–69.

[3] Chernyakhovskiy, I., et al., "Grid-Friendly Renewable Energy, Solar and Wind Participation in Automatic Generation Control Systems," *National Renewable Energy Laboratory*, 2019.

[4] PJM-ISO, "PJM Manual 12: Balancing Operations Revision 43," PJM.com, June 6, 2021.

[5] Hsia, E., "Energy Storage in PJM: A Perspective," PJM ISO, September 16, 2019, https://insidelines.pjm.com/energy-storage-in-pjm-a-perspective/.

[6] Skrzypczak, J., and B. Enayati, "Advanced Use Cases of Batteries on Nantucket," June 26, 2020, https://www.tdworld.com/distributed-energy-resources/energy-storage/article/21132541/national-grid.

[7] Gheorghiu, I., "There Once Was a 48 MWh Tesla Battery on Nantucket, Which Saved National Grid $120M in its Budget," October 10, 2019, https://www.utilitydive.com/news/Tesla-national-grid-battery-energy-storage-8hour-long-duration-diesel-generation-system-nantucket/564428/.

[8] Key, T., "IEEE Standard 1547(TM) Power Quality Considerations for Distributed Energy Resources," *EPRI*, Palo Alto, CA 2018.

[9] Ellis, T., "GVEA Fires Up Backup Generators After Earthquake Knocks Intertie Offline in Mat-Su," www.KUAC.org, December 4, 2018.

[10] DeVries, T., "Worlds Most Powerful BESS is Online and Working in Alaska," Golden Valley Electric Association, Fairbanks, AK, 2005.

[11] GVEA, "Battery Energy Storage System (BESS)," 2020, https://gvea.com/battery-system/.

[12] Environmental Protection Agency, "Sources of Greenhouse Gas Emissions," 2019, https://www.epa.gov/ghgemissions/sources-greenhouse-gas-emissions.

[13] "Storage Futures Study," National Renewable Energy Laboratory, 2022, https://www.nrel.gov/analysis/storage-futures.html.

[14] Jorgenson, J., et al., "Grid Operational Impacts of Widespread Storage Deployment," National Renewable Energy Laboratory, Golden, CO, 2022.

[15] IRENA, "Renewable Power Generation Costs in 2020," IRENA.org, https://www.irena.org/publications/2021/Jun/Renewable-Power-Costs-in-2020, 2021.

[16] Cole, W., and A. W. Frazier, "Cost Projections for Utility-Scale Battery Storage: 2020 Update," National Renewable Energy Laboratory, NREL/TP-6A20-75385, 2020.

[17] Denholm, P., et al., "The Four Phases of Storage Deployment: A Framework for the Expanding Role of Storage in the U.S. Power System," National Renewable Energy Laboratory, Golden, CO, 2021.

[18] Edison, T. A., "Apparatus for Translating Electric Currents from High to Low Tension," United States Patent #278,418, May 29, 1883.

Appendix 2A

AC versus DC

In the battle in the late 1800s over whether electricity was to be distributed in the form of ac or dc (for several reasons ranging from available technologies, politics, personalities, and, not to mention, corporate profits), ac won out. Consequently, most of all electricity distributed in the world oscillates at either 50 or 60 Hz (some shipboard power is 440 Hz).

Core to the issue of whether ac or dc is the most appropriate form of distributed electricity is the economics of transmitting of power over long distances. Edison's dc distribution systems were limited in their size because of the amount of copper that would be required to carry the low-voltage, high-current power. High-voltage, lower-current small wires were key to the economical wide-spread transmission of electricity. When Edison and Westinghouse were battling in the late 1800s over ac vs dc, there was no easy way to increase and decrease the voltage on a dc distribution wire. Edison patented a method [18] but it required two spinning machines coupled together on their axles. One being powered by the incoming dc voltage, and the other generating another dc voltage on its output. The ratio of dc voltages was simply a matter of the number of turns of wire in the coils of the respective motors and generators. But the complexity, size, inefficiency, and maintenance of the dual machines for this task made it uneconomical for widespread adoption. On an ac system, voltages can be converted using a simple power transformer. A step-up transformer at the generating station steps up the voltage to a higher level, and step-down transformers at the loads convert the higher voltages to a practical, useable, and most importantly, safe voltage that can be used to power consumer appliances. The elegant simplicity of a transformer contrasted with Edison's dual-coupled dc voltage machine was probably the single largest factor in why ac won out over dc to distribute electricity. Today, dc voltage conversion is performed electronically and allows efficient high voltage dc coupling between grids and over long distances. The transition to a fully dc grid is the subject of much debate at all levels.

CHAPTER 3

Mechanical Energy Storage

Every week, I stop at the grandfather clock in the hallway and pull a chain to raise a heavy counterweight up 1 meter. During the week, this weight tugs at a series of gears and pushes with a small force on an escapement gear that is allowed to rotate a fraction of a turn by a pivoting anchor mechanism, propelled by the escapement gear's push forward. As the pivoting anchor rocks back and forth, it gently nudges a long pendulum to swing side to side. With each swing, the heavy counterweight falls 1.5 μm toward the bottom of the clock's enclosure. The wound spring in other clocks does what the grandfather's gravity-driven weight does, but in a much more compact, space-efficient manner. These examples have energy stored in a mechanical form that is converted to kinetic energy to do something useful.

Mechanical energy can be of two forms. One is kinetic energy, such as in a flywheel where the energy is stored in the motion of a mass, quantified by $E_{kinetic} = 1/2 \cdot mv^2$. The other is potential energy, where the energy is stored in a raised mass, quantified by $E_{pot} = mgh$, or a wound spring, quantified by $E_{pot} = 1/2 \cdot kx^2$, that has the potential to do work if it were allowed to be released over a defined distance. This chapter will discuss the first two of these forms and give specific examples of each, along with their limitations and benefits.

3.1 Kinetic Energy Storage

Chapter 2 detailed the inertial energy stored in the mass of spinning, grid-connected generators, and described how this inertia helps to stabilize the grid in response to rapid fluctuations in loads and generation. The amount of this inertial energy stored in those spinning masses at any given moment is enormous, yet it would only provide a few seconds of power if it were suddenly the only source of power injected into the grid.

3.1.1 The Fundamentals

What if inertial energy storage was taken to a whole new level and could store enough energy to deliver power for minutes or hours, instead of seconds? To see if this is possible, let us review the basic principles of inertial energy. The energy in a spinning mass is determined by:

$$E_{rot} = \frac{1}{2} J\omega^2 \tag{3.1}$$

Where J is the moment of inertia, a constant specific to the geometric shape, weight, and size of the spinning mass. The larger the physical size and weight of the spinning mass, the greater the amount of energy that can be stored for a given rotational velocity, ω. Note that the energy stored in a spinning mass is proportional to the square of its rotational velocity. Consider the following hypothetical example. Assume a cylinder about the size of an 18650 battery cell (Figure 3.1) is spun about its z-axis at 10,000 rotations per minute (RPM), or 167 per second.

The moment of inertia of a cylinder is:

$$J = \frac{1}{2} M \cdot R^2 \tag{3.2}$$

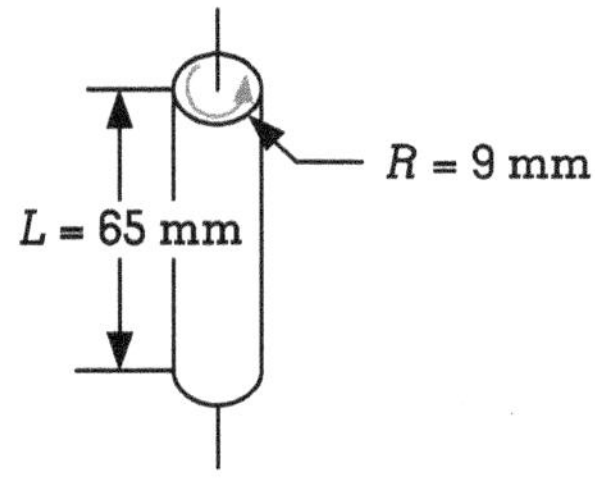

Figure 3.1 Hypothetical 18650 cylinder used as a flywheel.

Where M = mass of the cylinder, and R is its radius.

Using (3.1) and (3.2) the energy in a spinning 18650 cell is:

$$E_{18650} = \frac{1}{4} M \cdot R^2 \cdot \omega^2 = \frac{1}{4} \cdot M \cdot 0.009^2 \cdot (2 \cdot \pi \cdot 167)^2$$
$$E_{18650} = 22\ (J/kg) = 0.005\ (Wh/kg)$$

If the mass of this 18650 battery cell is 0.05 kg, then its volumetric energy density as a flywheel would be 0.02 Wh/liter. Contrast that to a lithium-ion cell which can store about 230 Wh/kg, or almost a megajoule of electrochemical energy in the same form factor. To make this spinning cylinder have as much energy as its electrochemical counterpart, it would have to spin at nearly 2 million RPM.

Admittedly, this is a poor representation of a flywheel. Flywheels with a greater radius-to-length ratio will store more energy for a given weight. In fact, the mass of a cylinder, M, is dependent on R^2:

$$M = \rho \cdot V = \rho \cdot \pi \cdot R^2 \cdot L \tag{3.3}$$

Where ρ = density, V = volume, and L = length of the cylinder.

Substituting (3.3) into (3.2) shows that the moment of inertia of a cylinder is proportional to R^4:

$$J = \frac{\rho \cdot \pi \cdot R^4 \cdot L}{2} \tag{3.4}$$

Therefore, for a given material density, a flywheel's potential to store energy is proportional to R^4 and only linearly proportional to its length, L. We will see some examples of better-designed, real-world flywheel energy storage systems later in this section.

Low energy density aside, there are some advantages to this form of energy storage. For one, since the mass and geometry never change, the intrinsic energy that it can store stays the same, unlike many electrochemical cells, in which the energy that they can store degrades over time and use. Naturally, the physical components, such as vacuum pumps, seals, electronic bearings, and cooling equipment are all subject to wear over time. However, a well-designed system will make the diagnosis, repair, and replacement of these items quick and easy to minimize the downtime of the energy storage system during its long service life.

Another advantage is their round-trip efficiency. Flywheels are spun by bidirectional motor-generators that use grid energy to motor the flywheel up to its operating speed. When needed, they then generate electricity from the inertial force of the spinning mass. Most of the round-trip losses occur in the wiring and electromagnetic interfaces in the motor-generator, as it converts electricity to mechanical energy and vice versa. Motor-generators can be made to be extremely efficient using precise coil designs and arrangements, high-quality magnetic materials, permanent magnets, and intelligent controls. Typically, one can expect to see about 2% to 5% one-way power losses through the motor-generator when it is in use.

Because its rotational speed will vary throughout its charge and discharge cycles, it cannot be directly synchronized to the ac grid. Typically, a variable-frequency, two-stage, power conversion system is used to control the flywheel's speed and interface it to the power grid. The first stage converts the grid's fixed frequency ac power to a higher dc voltage. A second stage converts the dc to a variable-frequency, three-phase ac voltage that can be used to either drive the motor to spin up the flywheel or accept electrical power from the flywheel as it ramps down. Using high-speed switching devices, the conversion stages operate at a much higher frequency than their input and output ac frequencies, allowing for a highly efficient conversion process. One can usually expect about 1% to 2% one-way losses in each of these power stages, while running at full power.

In addition to the electrical losses, frictional forces will also contribute to a continuous loss of state-of-charge. In order to compensate for these losses, a small amount of power is required from the electrical supply, reducing the overall round-trip efficiency of the system.

One source of friction comes from the outer surface of the flywheel rubbing against air molecules. Although it seems like a minor force, when the flywheel is spinning at a very high speed, that friction can cause significant drag. For example, the outer circumference of a flywheel with a diameter of 1 meter, spinning at 10,000 RPM, will travel at 1,900 km/hr! Therefore, most flywheel designs will incorporate a sealed vacuum chamber in which the flywheel is spun. The vacuum pumps will occasionally operate to maintain the chamber's low internal pressure, but that is a small amount of power compared to that which would otherwise be lost due to air drag on the flywheel surface.

Another source of friction comes from the axle bearings. The typical grease-packed bearings of ordinary spinning machines will cause too much drag in high-speed flywheels. Typically, flywheels incorporate a magnetic bearing where the axle is levitated away from its supports and floats on a cushion of magnetic force, virtually eliminating the friction between the moving surfaces. Some magnetic bearings use superstrong neodymium rare-earth magnets to levitate the shaft or reduce its downward force on a mechanical bearing. Electronically controlled electromagnets are also widely used, where an electric current is driven through a series of coils that surround a steel shaft and is controlled so that the shaft is perfectly centered inside the bearing. If any wobble occurs because of outside disturbances, the controller adjusts the current in one or more coils to return the shaft to equilibrium. Superconducting electromagnets are beginning to find more use in this application. Unlike ordinary electromagnets, in which the steady flow of current through their coils causes them to heat up and deducts from the overall efficiency of the flywheel system, superconducting electromagnets have no coil resistance. They may be electrically more efficient but require extremely low-temperature chillers to maintain superconductivity around the coils.

The power required to drive the magnetic bearings, vacuum pumps, and chillers, will cause the flywheel to lose energy over time. Unlike a battery, which can be charged and left on the shelf, unattended for months or even years, a flywheel will steadily lose its energy in a matter of hours or days. Typically, about 3% to 10% of the energy stored in a flywheel is used to power the electromagnets and/or chillers on an hourly basis.

There are several different approaches to building flywheels in the industry today. Some designs include heavy, solid steel rotors, which are spun at a low speed (under 10,000 RPM). Others use superstrong, lighter-weight, high-speed rotors that can spin well over 10,000 RPM. At these speeds, the challenge is that the forces in the rotating disk are tremendous. Advanced materials, such as carbon fiber, are used to hold the disk together against the immense centripetal forces.

3.1.2 Beacon Power

In the early 2000s, a company called Beacon Power (spun off from SatCon) created a high-speed flywheel energy storage system which was mainly used for high-power, high-cycle, low-energy frequency regulation in the eastern United States. The spinning rotor is a carbon fiber reinforced hollow cylinder, to give it strength to withstand years of high-speed rotations at around 16,000 RPM. Inside the hollow cylinder, are the motor-generator and magnetic lift system, the latter of which lifts the weight of the flywheel off the bearings to eliminate frictional forces. The energy required to operate the bearings, vacuum pump, and controls for 1 hour, is about 3% of the total stored energy in the wheel when fully charged.

The entire unit is encased in a vacuum chamber and installed in a buried, concrete-reinforced silo below ground (Figure 3.2). The diameter of the flywheel's vacuum chamber is 120 cm while its height is 200 cm. It can store about 36 kWh of useable energy, making its energy density about 16 Wh/l. The power systems that drive the motor and generate power are sized so that they can spin up the rotor in 2 to 3 minutes and reconvert that energy back to electrical energy in 2 to 3 minutes. Beacon Power claims that its flywheels can do this an unlimited number of times in a 25-year service life.

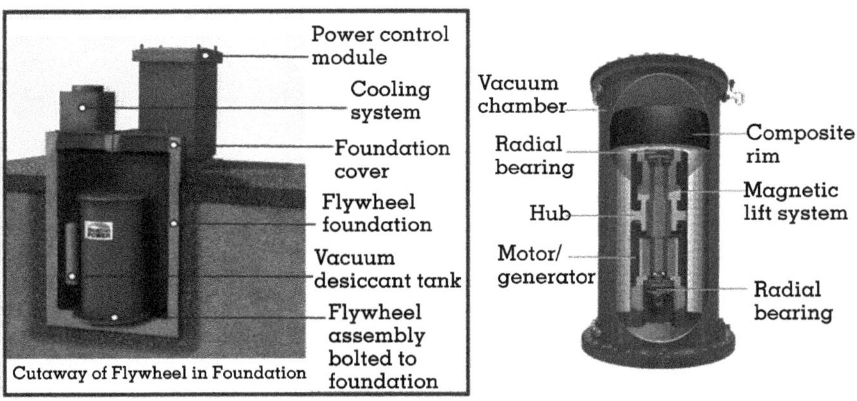

Figure 3.2 Beacon Power's medium speed, flywheel energy storage system. (Source: Beacon Power. Reprinted with permission.)

The Beacon Power product is well suited for high-power, low-energy applications, such as frequency regulation, frequency response, ramp-rate control, and artificial inertial response. Its high cycle-life, relatively high efficiency, and fast response time allow it to operate for 25 years with extremely high performance metrics that do not degrade over time. This is an extremely important point when evaluating technologies for energy storage applications. The number of cycles that a flywheel can execute over its lifetime is far greater than most electrochemical batteries. Therefore, when comparing the cost of a flywheel with that of a battery in high-cycle applications, the cost per cycle is an important attribute.

Beacon Power installed two significant frequency regulation systems in the United States that are still operating today. Each one is installed in a different ISO region, such as New York ISO (NYISO) and PJM.

Beacon Power's first 20-MW installation is in Stephentown, New York, where it has been operating since 2011 (Figure 3.3). Its 200 flywheels cycle many times each day, every day of the year, providing frequency regulation into the NYISO. Because it provides a very precise response to the ISO's power demands, it delivers almost 30% of the ISO's frequency correction services.

Subsequently, Beacon Power installed another 20-MW facility in Hazle Township, Pennsylvania, which has been in operation serving the PJM frequency market since 2014 (Figure 3.4).

Beacon Power also had a smaller 500-kW installation in the New England ISO (ISO-NE) region, which has been operating intermittently since 2008.

3.1.3 Flywheels for Frequency Regulation

Hypothetically, if flywheels were used in an application that required 100,000 cycles during a 25-year period, they would never have to be replaced. Conversely, if a competing lithium-ion battery were used for the same application, chances are it would have to be replaced five or more times during that same service life. This may seem like an exaggerated number of cycles, equal to nearly 14 cycles per day, but this is consistent with typical frequency regulation profiles.

For example, the PJM frequency regulation requires a tremendous amount of throughput from an energy storage system. Throughput is the amount of energy going in and out of storage. A common metric used

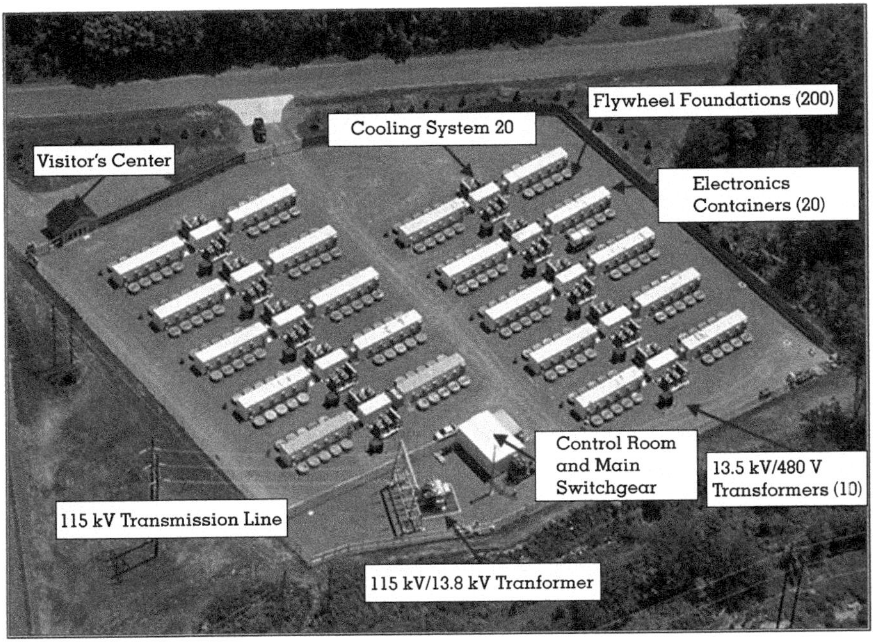

Figure 3.3 Beacon Power's frequency regulation site in NYISO. (Source: Beacon Power. Reprinted with permission.)

Figure 3.4 Beacon Power's frequency regulation site in PJM. (Source: Beacon Power. Reprinted with permission.)

to compare regulation protocols is megawatt-hour throughput (MWHTP) per MW per day. In 2019, the average throughput for PJM was 12.6 MWHTP/MW/day. If an energy storage system has 15 minutes of energy capacity, the total number of equivalent round-trip cycles in each day will be 25. Figure 3.5 shows the summary stats of the signal sent to the PJM FR assets for each day in 2019.

The lowest throughput day is Christmas, December 25, while the highest is on November 13. Figure 3.6 shows the signal sent to the FR assets, and the resulting SOC of a 15-minute storage system. The flywheels perform very well in this application, following its signal very closely.

Figure 3.7 is a zoomed-in section between 5:45 AM and 6:45 AM, showing the signal (labeled "AGC"), the asset's response (labeled "Unit resp"), and the SOC. This particular period is highlighted because it shows the impact of the SOC on an asset's ability to follow the signal. At the top and bottom of the SOC range, the asset's response ("Unit Resp") is curtailed and no longer able to follow the desired signal (AGC). Over the whole day, the SOC mostly stays within the bounds and the performance score is in the high 90s, but some parts as shown are not compliant. A larger energy storage system with a longer run time will have a much higher performance score. Figure 3.8 shows how a 1-hour energy storage system responds, having a smaller SOC range, and is fully compliant throughout this period.

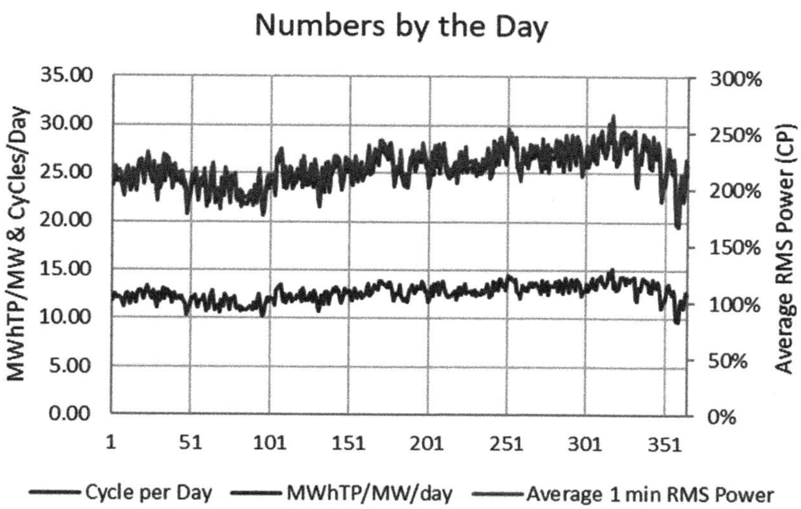

Figure 3.5 PJM frequency regulation statistics for each day in 2019 for a 15-minute energy storage system.

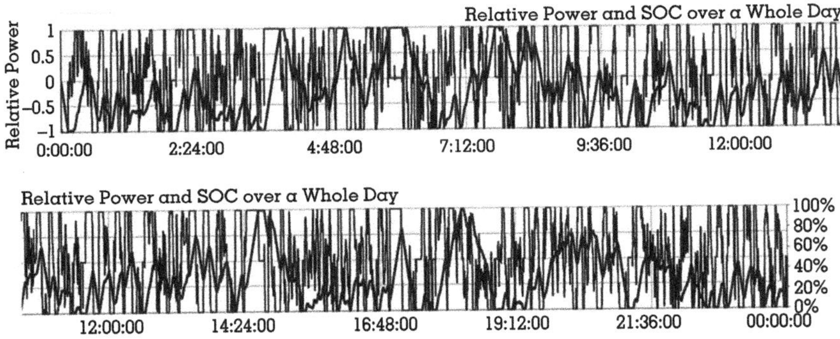

Figure 3.6 PJM FR signal and SOC of flywheels on November 13, 2019.

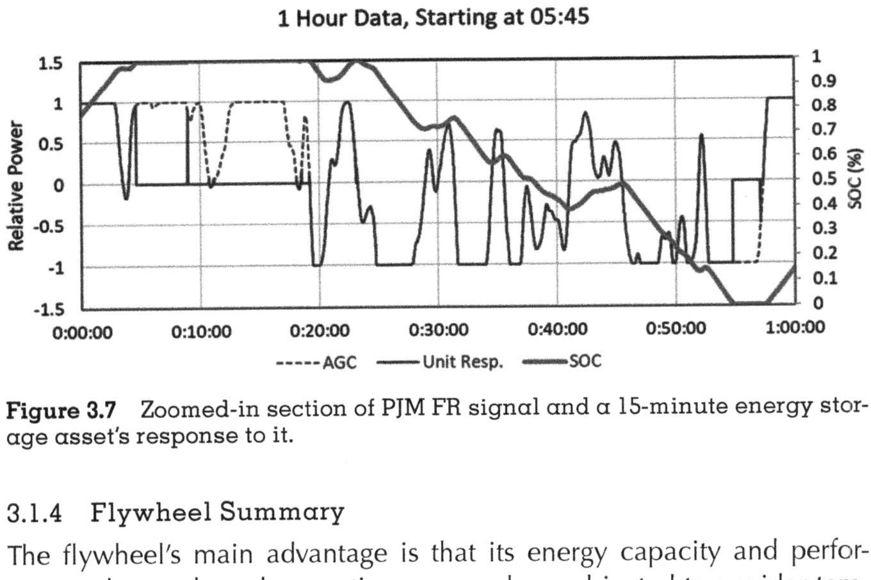

Figure 3.7 Zoomed-in section of PJM FR signal and a 15-minute energy storage asset's response to it.

3.1.4 Flywheel Summary

The flywheel's main advantage is that its energy capacity and performance do not degrade over time, even when subjected to a wider temperature range than which batteries are comfortable. They are also made of materials that are easy to source, such as steel, carbon, copper, and other abundant metals. Many manufacturers tout that they are environmentally friendly and sustainable, although some contain super-magnets made of neodymium. The round-trip efficiency of most flywheels is between 85% and 95%.

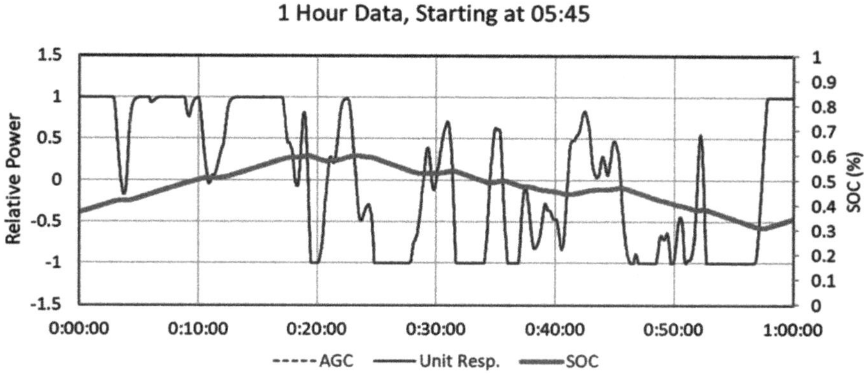

Figure 3.8 Zoomed-in section of PJM FR signal and a 1-hour energy storage asset's response to it.

Flywheels can store around 13 to 16 Wh/l but, when installed in an underground concrete enclosure, take up considerably more space. Because of their low energy density, many manufacturers have designed them for higher power applications, such as frequency regulation and inertial response services. Table 3.1 is a summary of the general characteristics of flywheel energy storage systems.

3.2 Potential Energy Storage

When raised to a higher elevation, an object[1] gains potential energy proportional to the raised distance, h, the mass of the object, m, and the Earth's gravitational acceleration, g.

$$E_{potntial} = m \cdot g \cdot h \tag{3.5}$$

The work being done on an object is equal to the force times distance; the work is the energy put into it. To make use of this work, the object will later be released from its hold, whereby most of its potential energy will be converted to kinetic energy that can be harnessed to drive a generator.

1. More accurately, the system including the object the environment interacts with increases its potential energy.

Table 3.1 Flywheel Summary Performance

Specification Parameter	Value	Notes
Energy density, volumetric	2–16 Wh/l	May be lower when accounting for installation in underground bunkers
Energy density, intrinsic	1–70 Wh/kg	—
Cycle life	Infinite	No degradation from cycles
Calendar life	20–30 years	Estimates from vendors, usually based on electric and electronic failure degradations
Round-trip efficiency	86% to 95%	—
Self-discharge	1% to 10% per hour	—
Operating temperature range	-40°C to 65°C	Limited by controls and product design, not anything inherent to the technology
Failure modes	Rapid physical disassembly can occur damaging surroundings Power supply failures Bearing failures	Some flywheels are therefore installed in underground containment vessels to contain catastrophic failure debris avoiding the possibility of human injury and property damage
Maintenance requirements	Vacuum pumps and motors, bearing assemblies, electronics components and assemblies, coolant, coolant pumps, and motors	—
Representative manufacturers	Beacon Power (Now RGA Investments)	www.beaconpower.com
	Oxto Energy	www.oxtoenergy.com
	Vycon Energy	www.vyconenergy.com
	Teraloop	www.teraloop.org

Thinking about the grandfather clock and its elegantly simple mechanism for storing energy, could it be possible to use such a principle to power other things, such as lights, electronics, fans, and other practical appliances?

The weight in the grandfather clock is about 6.6 lb (3 kg). It has a full range of motion of about 1m from top to bottom. At the top of its range, it has

$$E_{weight} = 3kg \cdot 9.8\frac{m}{s^2} \cdot 1m = 29J$$

of energy. Since a joule is a watt-second, this would provide a 10-watt light bulb with about 3 seconds worth of power. In another example, an 18650 cell weighs 50g and, as a battery, it can store about 230 Wh/kg of electrochemical energy. To determine how high it would have to be raised to achieve an equivalent amount of potential energy, set 230 Wh/kg = *mgh*/kg, and solve for the required height:

$$230\frac{Wh}{kg} \cdot 3600\frac{J}{Wh} = 9.8\frac{m}{s^2} \cdot h$$
$$h = 84.4km$$

Using spent batteries as dead weight in a gravity-based storage system would require a lot of batteries and height to store a useful amount of energy. However, that has not stopped people from using gravity to store energy in other ways. For one, hydroelectric energy storage is a form of gravity-based energy storage in which water is held at a raised elevation and released when energy is needed (more on this in Chapter 4).

A few companies are researching other methods of gravity-based energy storage. Gravitricity is looking at elevating and lowering heavy weights in a deep mine shaft using motors and cables. Energy Vault has designed and is building a stack of dense blocks using a crane or a structure of elevators. Gravity Power is seeking to pump water beneath a giant piston in a deep shaft to lift a massive piston. All of these companies are using massive weights to store potential energy in useful quantities. Although these are not very energy dense, they could be useful in a stationary grid application where size and weight are not as important.

3.2.1 Gravitricity's Weights on a Cable

Gravitricity's initial concept has a motor that turns a winch that pulls on one or more cables tied to a pair of 25 metric tonne weights (Figure 3.9). The potential energy associated with such a weight can be calculated with respect to its raised height as:

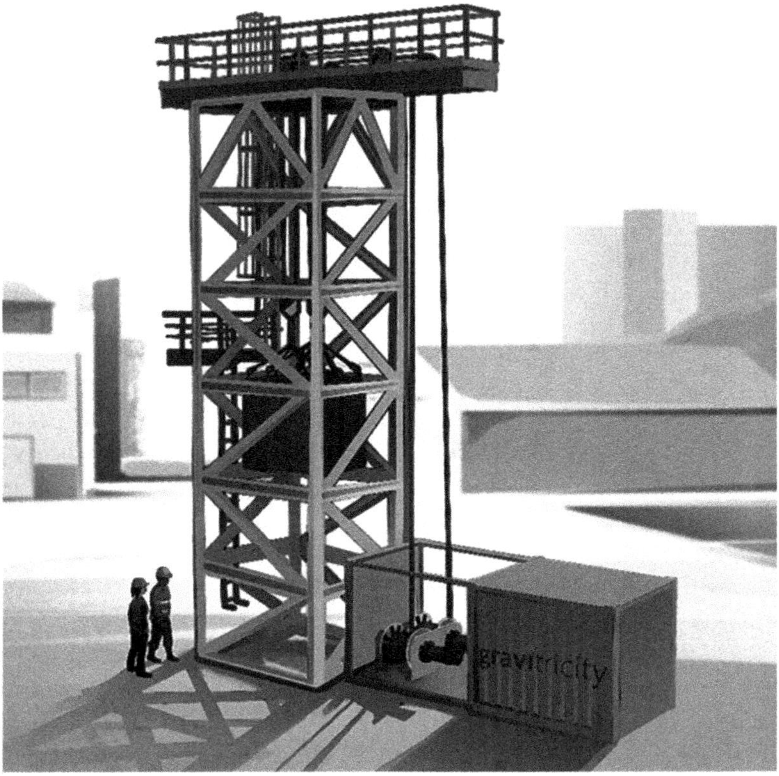

Figure 3.9 Gravitricity's 250-kW test rig in Leith—artist's impression (Source: Gravitricity. Reprinted with permission.)

$$E_{potential} = mgh \text{ (Joule)} = \frac{50{,}000kg \cdot 9.8\frac{m}{s^2}}{3600 \text{ (J/Wh)}} = 136 \left(\frac{\text{Wh}}{\text{m}}\right)$$

In their prototype system set up on the docks of Port of Leith, Edinburgh, Scotland, their two 25-tonne[2] weights, moving up 15m, can store 2 kWh. If their motor can put 250 kW into the system, the 50-tonne weight would move up at about a half meter per second, traversing the entire span of 15m in just 30 seconds. To release the energy, they switch

2. 1 tonne = 1,000 kg.

the motor to a generator with precise electrical control of the current in the motor's windings. They then release the brakes and allow the weights to push on the generator, thereby supplying a constant electrical power into the power grid.

In the future, they envision these weights in deep mine shafts up to 1.5-km deep (Figure 3.10).

A 50-tonne weight in a 1.5-km shaft could store 200 kWh, and a 250-kW motor would bring the weight up the length of the shaft in a little over 45 minutes. They have considered using weights between 500 and 5,000 tonnes, but the extra weight comes at a cost of heavier cables, increased losses in the gearboxes, and other system strains. A 5,000-tonne weight in a 1.5-km shaft would store the following energy:

$$E_{potential} = 5E6kg \cdot 1.5E3m \cdot 9.8\frac{m}{s^s} \cdot \frac{MWh}{3.6E9J} \cong 20MWh$$

Figure 3.10 Concept drawing of a deep mineshaft energy storage system (Source: Gravitricity. Reprinted with permission.)

If 5,000,000 kgs are raised in a shaft with four winches, each coupled to 10 × 1 inch steel cables, the total stretch at the bottom of the shaft would be 50m. The cables would weigh approximately 165,000 kgs, providing an additional 340 kWh of potential energy. There would be four spools of cable in the surface enclosure, each around 2m to 3m in width and diameter. The cable stretch is also a source of potential energy, like in a wound spring. The 50m of stretch times 5,000,000 kgs of force is approximately another 70 kWh of energy storage. When accounting for the weight of the cables and their spring tension fully stretched, the 5,000,000-kg weight over a 1.5-km drop can store up to 20.4 MWh of energy.

One of their more recent ideas is to have a mechanism that moves some of the blocks off to the side when they get to the top of the shaft; achieving a higher energy content without having to increase the physical strain on the winches. The mechanism to move the blocks on and off the lifting platform adds to the complexity of this once-simple idea but, if done well, could provide increased scalability to improve their metrics.

One of several advantages of this technology is that the fundamental energy storage medium never loses its capacity to store energy. Like a flywheel, the amount of energy never decreases because it is based on gravity and mass. Cables, bearings, and motors may wear, but the mass of the weights never decreases. If maintained properly, such a system can continue to operate indefinitely.

Another advantage is its efficiency since the only losses are the electrical losses in the motor-generator and those due to miscellaneous mechanical friction throughout this predominantly mechanical system. The expected round-trip efficiency would be about 90% for MW-sized systems.

One disadvantage of this technology is its potential construction costs. Digging a 1.5-km deep shaft is not a trivial task and would raise the up-front cost of this technology if required. Gravitricity suggests using abandoned mine shafts to help reduce costs.

3.2.2 Gravity Power's Floating Piston

Gravity Power is another company developing gravity-based energy storage with heavy weights but, instead of lifting the weight with cables, they pump water beneath a heavy piston with sealed edges, situated in a tall, vertical, smooth-walled shaft (Figure 3.11). Electricity is used to pump water from above to below the piston. The weight of the piston minus that of the displaced water is the net mass being lifted up the height of the shaft.

When discharging, the piston's weight forces water back up through turbines, which turns generators that push power into the grid. If the cylinder is made of concrete and stones, its density will be around 2,400 kg/m^3 compared with that of water at 1,000 kg/m^3. If a 30m round shaft is 155m deep and an 80m long cylinder slides up and down in it, the weight of the cylinder will be 136 million kg and displaces 57 million kg

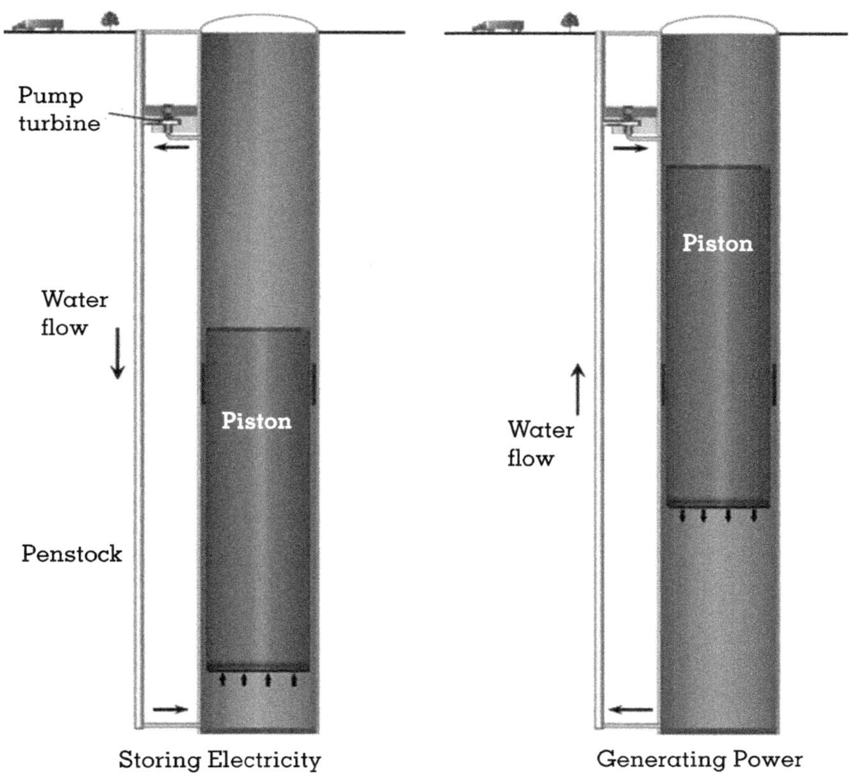

Figure 3.11 Gravity Power's concept of a large subterranean piston in a water-filled vertical shaft. (Source: Gravity Power. Reprinted with permission.)

of water. In the 155m long shaft, it can move up and down 75m, so the potential energy it can store is:

$$E_{pot} = (136E6 \text{ kg} - 57E6 \text{ kg}) \cdot 75m \cdot 9.8\frac{m}{s^2} \cdot \frac{MWh}{3.6E9J} = 16MWh$$

That is its ideal energy. Using water to move the piston, such a system will lose about 20% of the total energy during a complete round-trip cycle due to the inherent inefficiencies of capturing energy from moving water.

3.2.3 Energy Vault's Stack of Blocks

The final example presented in this chapter is the idea of stacking blocks to store electrical energy and, when electricity is needed, dropping them back down to a lower level. Energy Vault, a company based out of Switzerland, has been developing this idea since 2017. They created a proof-of-concept which uses a multiarmed crane (Figure 3.12) to lift and drop blocks in an orderly fashion to store and deliver electrical energy. Each block weighs 35,000 kg and has features on it allowing an autonomous crane to grab it and stack it on top of another block. The blocks are made from a proprietary composite material that they claim will last a lifetime and is sourced from local soil, coal ash, recycled glass fibers, and refuse material that would otherwise be thrown into a landfill.

Their original concept of a 150m tower is constructed by stacking blocks 39 layers high, 200 blocks per layer. The total mass is about 273,000,000 kg and, when stacked at an average of 75m high, represents about 56 MWh of intrinsic energy. The unstacked tower is around 12 layers high or an average of 23m high, leaving 38 MWh of achievable energy storage.

Like other mechanical energy storage concepts, this one is also advantageous for its longevity. With routine maintenance, these mechanical systems can potentially last forever. A challenge of this original concept is developing controls that can autonomously grab blocks and precisely place them on the next layer in the tower. This placement needs to be done quickly and autonomously, with well-equipped vision and

Figure 3.12 Energy Vault's concept of stacking and unstacking 35-tonne blocks to capture potential energy. (Source: Energy Vault. Reprinted with permission.)

other sensing devices to consistently place them where they need to go, no matter the weather conditions.

Energy Vault's most recent concept is a metal structure with horizontal storage rails and a lifting mechanism to lift blocks from one elevation to the next (Figure 3.13). The blocks are stored in rows on different levels in the structure. To add energy to the structure, blocks are moved to the edge of the row, where an elevator lifts them up to a higher level, then they are moved and arranged compactly along the higher row. When energy is needed, the blocks are lowered to a lower level by a similar process, but in reverse. The structure has multiple parallel elevators which can be engaged in independent movements at all times, thus allowing this system to rapidly respond to grid power requests. This concept is much more stable and predictable during high wind activity and earthquakes than their stack of blocks concepts. However, it will have an inherently higher installation and materials cost due to the added structure and mechanics compared with the simple crane idea.

Figure 3.13 Energy Vault's new building-based gravity energy storage system. (Source: Energy Vault. Reprinted with permission.)

3.2.4 Summary of Potential Energy Storage Systems

This chapter presents a few of the many potential energy concepts and products being explored by companies around the world. New ideas for gravity-induced potential energy continue to show up in news feeds, including undersea balloons tethered via cables to sea-bottom motor-generators, helium balloons tied to terrestrial machinery, grid-interactive elevators, train cars on hills, and more. The common theme among them is their relatively high efficiency and lifetime longevity. On the downside, these technologies can consume a lot of volume and, therefore, it can be challenging to find the physical space where they would be electrically useful. Location issues aside, they could serve a wide variety of grid applications from frequency regulation to energy shifting, using electronically controlled, variable-speed motor-generators for fast and efficient grid interactions. Table 3.2 is a summary of the general characteristics of potential energy storage systems.

3.3 Mechanical Energy Storage Summary

Mechanical energy storage has been around for thousands of years. Ever since the potter's wheel, heavy weights have been used to smooth the operation of our daily activities. Potter's wheels, spinning wheels, water

Table 3.2 Summary of Potential Energy Storage Systems

Specification Parameters	Values	Notes
Energy density—volumetric	Varies	Extremely low, but sometimes takes up relatively no surface space as the storage is buried in the earth
Energy density—intrinsic	Extremely low	In a 1.5-km shaft, Gravitricity's weights are 4 Wh/kg. In Energy Vault's 150m tower, their 35-tonne blocks are 0.14 Wh/kg.
Cycle life	Infinite	No limits
Calendar life	Infinite	No limits with proper maintenance and parts replacements
Round-trip efficiency	80% to 90%	Larger motors and generators will have higher efficiencies.
Operating temperature range	Very wide	Not limited by batteries, only power electronics and rotating EM machines
Failure modes	Structural collapse, cable failure, pump loss, electronics failure, and bearing freeze	—
Maintenance requirements	Mechanical and electrical components inspections, repairs, or replacements	—
Representative manufacturers	Energy Vault	https://energyvault.com/
	Gravitricity	https://gravitricity.com/
	Gravity Power	www.gravitypower.net

wheels, gristmill stones, and steam engine flywheels are examples of kinetic energy storage. Mechanical clocks, bows and arrows, slingshots, and catapults are examples of stored potential energy. For years, theme parks have used gravity-based energy storage to launch their happy customers around a gut-wrenching array of turns, twists, hills, valleys, and gravity-defying loop-de-loops.

Recently, attention turned to using the laws of motion and bending the space-time continuum to store energy and smooth the operation of our daily electricity production. Despite its large up-front costs and size of the plant in which the energy is stored, the storage medium for this type of energy storage never degrades. Only the mechanisms that move, protect, and interface with the medium may wear and need replacement over time. This aspect makes it potentially the cheapest energy storage ever with respect to the cycled energy it delivers over its entire service life.

Additional References and Useful Links

https://www.explainthatstuff.com/flywheels.html

CHAPTER 4

Hydroelectric Energy Storage

Hydroelectric energy storage, also known as pumped hydro storage (PHS), has been around for over a hundred years since the Swiss first started using it to smooth the power output from their river-derived power plants into their thriving industrial centers. It quickly spread throughout Europe but took a few decades to be adopted in the United States and other parts of the world. Once in the United States, it took off, with the country building hundreds of gigawatt-hours of energy storage from the 1950s until the early 2000s. Worldwide, PHS is still the dominant form of grid-based energy storage, providing around 95% of all energy storage resources that help manage the varying grid load between day and night. Although PHS was initially built to help the nuclear and coal power plants run at steady outputs, it now helps to compensate for the variable power production of renewable energy resources.

4.1 The Basics

PHS relies on geography that allows a massive amount of water to be pumped from one elevation to another. Huge volumes of water provide the mass that is used to convert electrical energy to potential energy and back again through a network of pumps, pipes, turbines, and motor-generators. The difference in height between the two water sources (e.g., reservoirs) directly affects the amount of energy that can be stored by the water in each. As Chapter 3 described, the work put into raising a mass to a higher elevation increases its potential energy, which is mostly converted to useable electrical energy upon falling back to its original height. With water, the ratio of output energy to input work is lower than other forms of mechanical energy because there are more losses in the process of pumping water and extracting energy from its motion. In general, for most operating PHS facilities, this ratio is about 70% to 80% round-trip depending on the design and age of the facility [1].

The left side of Figure 4.1 illustrates the electrical grid powering a motor that pumps water from a lower water source to an upper reservoir. That water flows back down through a turbine generator, shown on the right side of the diagram, to produce electricity that flows back into the grid. Although the diagram shows a separate pump and turbine generator, most installations use a dual-purpose machine, controlled as a motor when pumping water uphill, and controlled as a generator when letting water flow back down.

Notably, some of the very first PHS systems used separate motors and pumps because this was easier to design at the time. As the technology progressed, engineers combined the machines, impellers, turbines, and pipes into one set to reduce the cost of the installation and operation. Figure 4.1 also shows the lower water reservoir as a river, which is often the case. However, the lower water source can also be a man-made reservoir, natural lake, or even the ocean. In all cases, a natural water source is required to make up for evaporation that occurs at the surfaces of any bodies of water exposed to the atmosphere. Evaporation is this energy storage technology's chief self-discharge mechanism.

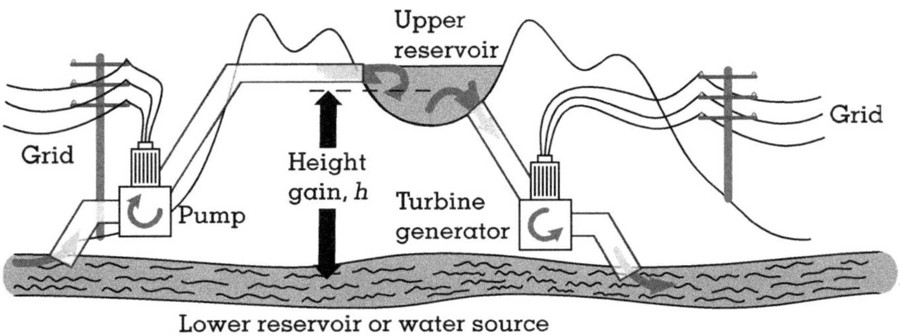

Figure 4.1 A drawing showing the basic process of hydroelectric energy storage.

4.2 Massive Scale

With any mechanical energy storage system, storing a useful amount of energy requires a significant amount of mass—PHS is no exception. To get a megawatt-hour of storage, 4 million liters of water (almost two Olympic-size swimming pools) would need to be pumped to a height of 100m. Finding the land to develop and radically transform into a man-made reservoir is difficult, but not impossible. Back in the mid-1900s, this was done more freely than it would be now, but it was still not without pain. Sometimes, PHS projects involved the complete upheaval of family homesteads, farms, and even whole towns [2]. If the areas were forested, the trees and major vegetation were removed to reduce biomaterial in the water. While in some cases, the new reservoir doubled as a recreation area and was welcomed by the local population, in others, there was a lot of resentment from those who had to be forcibly moved, some from their family's ancestral and/or tribal lands.

4.3 Environmental Impact

When operating a PHS facility, the water levels and rate at which the water is pumped and dumped must be controlled to ensure the safety of the people and surrounding environment. Around 1988, Boyd Kynard, a migratory fish expert, was bass fishing in a small aluminum boat, 2 km south of the suction pipes of the Northfield Mountain PHS facility. Sitting quietly in this 160m-wide and slow-moving section of the river, he suddenly observed his boat being drawn upstream toward the PHS inlets "at a pretty good clip."[1] [3]. Likewise, when too much water is dumped into a public body of water, it can cause a dangerous surge of water that imperils boaters, swimmers, and animals downstream. Therefore, the rapid change of a PHS's energy (i.e., the power input or output), needs to be controlled and timed to manage the safety and environmental impact it has on its surroundings. Boyd Kynard reports that even subtle tidal-like patterns induced by the daily cycling of these facilities can have a detrimental effect on local wildlife not adapted to tidal conditions.

1. Kynard said that the only consequence for him that day was that he was unable to continue fishing for bass using the bottom-grub bait he intended to use, but the consequence to the environment is much more dire.

There are two basic types of PHS: open-loop and closed-loop. In an open-loop PHS, one or both reservoirs are part of an existing natural water system, such as a lake, river, or ocean. A closed-loop PHS system consists of two man-made structures and has little interaction with the local environment. The former system requires extensive environmental analysis, review, and approval to ensure that its operation does not negatively impact the environment, wildlife, and local populations. The latter system is easier to permit because the bodies of water are intentionally isolated from local wildlife and people.[2] Additionally, because of their isolation, closed-loop PHS system's water can be freely pumped between reservoirs in whatever quantities and rates the system is capable of handling without affecting people, wildlife, or the environment. The downsides of a closed-loop system are that both reservoirs need to be constructed and then securely guarded to prevent unsafe interactions with people and wildlife. Furthermore, the reservoirs need to be constantly replenished by a local water source to combat evaporation. One form of closed-loop PHS being contemplated uses hollowed-out caverns, tunnels, and wells of abandoned mine sites. Many unused subterranean mines could store millions of gallons of water, often at tremendous depths below the surface.

4.4 The Muscle in the Works

The turbine is a critical component of PHS, responsible for both converting waterpower to electricity and pumping. Historically, there are three main types of turbines used in hydroelectric energy generation: Pelton, Kaplan, and Francis.

1. The Pelton turbine shown in Figure 4.2 [4], was invented in the 1870s by Lester Allan Pelton, to capture the energy from small, swift-flowing streams in order to support the California gold-mining efforts. Turbines based on this concept excel in applications where the ratio of head (the height of the falling water) to flow is high, because they are very good at capturing the kinetic energy from fast-moving water. There are many variations of his

2. Some closed-loop systems may be isolated from natural bodies of water, but still open for public recreational use, although most are isolated from both.

Figure 4.2 Pelton Turbine has scoops to capture the energy from high-pressure, low-volume water streams. (Source: Dingley, A.,Wikimedia Commons [4].)

general design, the best having achieved almost 90% efficiency. The largest Pelton turbines in the world are the three 432-MW turbines in service at the Grande Dixence hydroelectric generating complex in Switzerland [5]. The dam itself is 285m high, holds back 400 million m^3 of water, and is 1,800m higher than its generating stations on the Rhine. This enormous head of water creates 185 bar of pressure driving the water into the turbines at 690 km/hr, which helps to efficiently generate 1.2 GW of power with a modest[3] flow rate of 75 m^3/s.

3. Modest compared to other generating stations with lower heads.

2. The Kaplan turbine shown in Figure 4.3 resembles a propeller that spins around an axis parallel to the flow of the water, with vanes that can be adjusted to regulate power and flow. It is named after Victor Kaplan, who patented the idea in 1912. Because of its adjustable vanes, it is adaptable to a wide range of water heights and flows. Today, it is used in many hydroelectric generating facilities where the height of the falling water is relatively low and the volume through them is high. Under ideal conditions, these turbines can achieve 90% efficiency.
3. The third type of turbine is the Francis turbine (Figure 4.4), named after James B. Francis, that works via a combination of axial and radial flow principles. This extremely versatile turbine can achieve efficiencies as high as 95%, is usable with heads between 10m and 300m, and can be used to pump water backwards by powering its reverse motion with a motor.

Figure 4.3 One of the Kaplan turbines from the Bonneville Dam after 61 years of service. (Source: Duk, Wikimedia Commons.)

Figure 4.4 The runner of the Francis-type turbine of the Imatrankoski power plant in Imatra, Finland, in use from 1936–1989. (Source: SeppVei, Wikimedia Commons.)

In the 1800s, several different versions of the vertical-axis mill were created by various inventors, culminating with the inward flow Francis turbine. Francis worked as head engineer of the Locks and Canals Company, located in Lowell, MA. He took the best features of existing vertical axis designs and created his own using advanced mathematical and empirical analytical techniques. Using a slide rule and graph paper, his estimation for the efficiency of his first full-scale turbine in 1849 was only off by 0.06%. Although his design was only incrementally better than any which preceded his, the real value he brought to the industry was the set of methods and analytical tools used to design a turbine that is optimal for the unique set of conditions associated with a facility. Many inventors contributed to further improvements of the Francis turbine over the years. Variations of his design are the most common form of generating turbines and are used almost exclusively for dual-purpose pump-turbines in PHS plants.

4.4.1 How Much Power Can a Turbine Produce?

Power from a turbine ultimately comes from the volume and force of water pushing the turbine blades, or fins, as they rotate around an axis. The more force and flow through the turbine, the more power is asserted on the turbine's axle. The generalized formula for this relationship is:

$$P = \rho \cdot \dot{Q} \cdot g \cdot h_u \cdot \eta_t \tag{4.1}$$

Where P is the mechanical power on the turbine's axle. Water density times the volume of water flowing in a given time, $\rho \cdot \dot{Q}$, represents the mass, m, of the water flowing through the turbine, dm/dt in units of kg per second. dm/dt, combined with the terms g and h_u, is $d(m \cdot g \cdot h_u)/dt$ which contains the formula for potential energy, $m \cdot g \cdot h_u$. Therefore, $d(m \cdot g \cdot h_u)/dt$ is the rate of change of the potential energy. The term for head, h_u, ideally would represent the height that the water falls in a frictionless world.

However, the friction and water turbulence in the penstocks, as well as through the various gates and channels, reduces the pressure on the turbine and contributes to the loss of converted energy. This is similar to how the resistance in a conductor causes the voltage seen by the load to be reduced in proportion to the resistance and current. Typically, a well-designed hydroelectric facility will see no more than 10% head loss through the water channels; therefore, the useable head, h_u, is usually 90% of the actual difference in height between the two bodies of water. Finally, the one-way turbine efficiency factor, η_t, is the ratio of incoming energy of the water at the turbine to the mechanical energy on the turbine's axle. This could vary from as much as 70% to 95%, depending on the design of the turbine. Equation (4.1) calculates the power produced by the falling water (i.e., the rate of flowing energy). To determine the energy produced by the fallen water, integrate (4.1) to obtain:

$$E = \rho \cdot V \cdot g \cdot h_u \cdot \eta_t = m \cdot g \cdot h_u \cdot \eta_t \tag{4.2}$$

Where m and V are the mass and volume, respectively, of water that flows over a given period.

4.4.2 Height and Flow

Equation (4.2) shows that just as in other gravity-based, mechanical energy storage technologies, the potential energy is proportional to the mass and height of the lifted weight. As an example, the Hoover Dam is 180m tall and is filled with water from the Colorado River (which flows around 600 m^3/s of water). When the dam is full, this continuous flow produces almost 1 GW of power. In contrast, the Grande Dixence hydroelectric system has a dammed reservoir about 1,800m above its generating station and can generate 1 GW with just 70 m^3/s of water flowing through its turbines.[4]

4.5 Examples of Pumped Hydro Storage

In the early 1960s, Tennessee Valley Authority (TVA) envisioned an energy storage facility situated on Raccoon Mountain in Tennessee [6, 7]. Although U.S. Congress members thought the idea was "insane," TVA prevailed and received funding for the project. Conceived in the early 1960s, construction on the facility did not begin until 1970, and it was substantially completed by 1978. The upper reservoir was hollowed out of the top of the mountain (Figure 4.5) and dammed with a 2,590 × 70m rock-filled dam; it can store 400 million m^3 of water, although it will only use a fraction of that amount during daily cycling.

The average height of this reservoir is 300m above the lower water source (Figure 4.6), the Nickajack Reservoir, on the Tennessee River. The facility consumes and delivers approximately 1.6 GW of power while filling and draining the reservoir. The upper reservoir is isolated from wildlife and recreation, allowing TVA operators the ability to completely drain it when necessary, which they do about every five years for inspections.

Although connected at the base directly to the Tennessee River, the lower reservoir is the Nickajack Reservoir, located a few miles downstream of the discharge ports. When discharging at full power, 440 m^3/s of water flows from the upper reservoir through a single penstock and is

4. Assuming that the conversion efficiencies of both sites are nearly the same. These figures are representative to show the effects of height and flow on energy production.

Figure 4.5 Part of the man-made upper reservoir of the Raccoon Mountain PHS facility; picture taken at a time of low water level. (Source: TVA.)

directed through four reversible pump-turbines in a hollowed-out cavern at the base of the mountain (Figures 4.7 and 4.8).

As large as this facility is, it is not the largest in the world. As of 2020, according to the U.S. DOE list of energy storage projects around the world [8], there were 14 larger operational PHS facilities located in the United States, China, Europe, India, and Japan. The 3-GW Bath County Pumped Storage Station in Virginia [9] has two reservoirs connected to two different creeks at an elevation difference of 380m. There are three enormous penstocks, each 9m in diameter and 1300m long, which pass water through six 500-MW pump-turbines. The total amount of useable energy that can be stored in this facility is 24 GWh. The reservoirs are closed to public access because their water levels rise and fall between 18m and 32m each day.

In early 2022, China State Grid announced the operation of the new world's largest PHS facility located in the Hebei Province. The Fengning Pumped Storage Power Station has twelve 300-MW generators

Figure 4.6 Raccoon Mountain man-made reservoir above the Tennessee River. (Imagery ©2022 Google, Imagery © 2022 Maxar Technologies, U.S. Geologic Survey, USDA Farm Service Agency, Map data © 2022.)

Figure 4.7 Hollowed-out cavern containing Raccoon Mountain's turbines and generating equipment. (Source: TVA.)

and is expected to cycle 18 GWh of electricity per day [10]. This is part of China's plans to have as much as 120 GW of PHS by 2030.

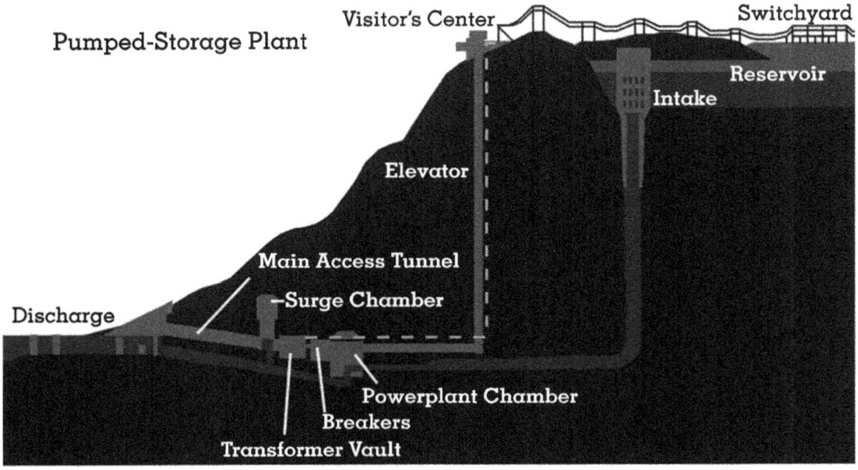

Figure 4.8 Cross-section diagram of Raccoon Mountain PHS facility. (Source: TVA.)

Additional examples of PHS facilities, 1 GW and larger, are listed in Table 4A.1.

4.6 PHS Applications

Although originally intended to provide bulk daily energy shifting, PHS is also capable of providing short-term power services and some seasonal energy storage, depending on the facility.

4.6.1 Bulk Energy Storage

Bespoke PHS plants, such as those listed in Table 4A.1, are designed specifically to move bulk energy from one time of the day to another and to occasionally balance weekly or monthly load variations. Although these facilities have their own reservoirs and are not as susceptible to low river flows as a dammed run-of-the-river facility might be, they are all subject to gradual loss of water by evaporation and seepage through the earth. Water can evaporate from a surface at a rate between 100 and 400 gallons per acre per hour, depending on the local temperature, humidity, and wind. In drought conditions, creeks and other natural water supplies may not provide enough water to keep the reservoirs full and thereby lose some of their capacity. A well-designed facility will have some buffer built into the storage capacity to absorb most of the seasonal variations of water levels, but some years may be worse than others.

One such year was 2012 in the U.S. southeast. The drought lowered Virginia's Bath County Pumped Storage Station's reservoir level and, on one very hot day, demanded so much electricity for air-conditioning, that the station's operators had to turn the pumps on to recover energy storage for the next day, while every other generating station in the region was generating at full power [9]. Although this was an anomaly that year, events like this will happen more frequently as the weather becomes more unpredictable.

4.6.2 Seasonal Energy Storage

Although pumped hydro storage is generally not built to provide seasonal energy storage, hydrogeneration facilities often have such an innate ability, whether intended or not. In a dammed river hydrogeneration facility, besides providing a large head to maximize the power output, the dam creates a buffer to help the system ride through seasonal variations in water supply and flow. For example, the Grande Dixence in Switzerland gathers the summer water melt from 35 nearby glaciers and uses that to generate electricity in the subsequent winter months. Lake Mead, the reservoir behind the massive Hoover Dam, took 6 years to fill and can yield many seasons of energy storage at maximum capacity.

Nevertheless, the long-standing drought in the last few years has significantly decreased the Colorado River's flow into Lake Mead, jeopardizing its ability to support local irrigation, electrical power generation, and seasonal energy storage. Figure 4.9 shows that the water levels have been falling for the past 20 years.

4.6.3 Short-Term Energy Storage and Grid Services

Chapter 2 detailed several short-term power and grid ancillary services that require a relatively quick response time and variable power inputs and outputs. Even though most PHS plants are designed to provide many hours of discharge capacity, they also can be (and have been) used for short-term power and grid services. In fact, both PHS and hydrogeneration facilities can be used for frequency regulation and response services. Some generation architectures are better suited to follow the fast-

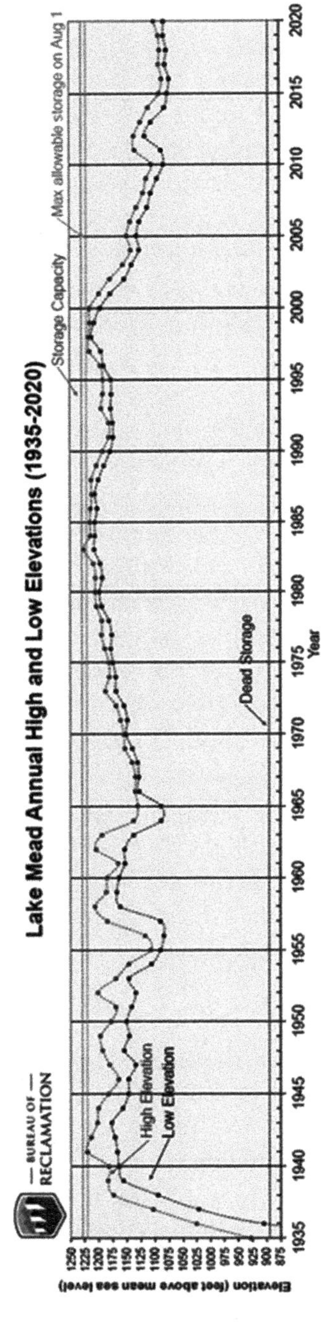

Figure 4.9 Record of water levels at the Hoover Dam since inception. (Source: https://www.usbr.gov/lc/region/g4000/lakemead_line.pd.)

moving regulation signals and primary reserve commands, while others are mainly used for slower secondary and tertiary responses. Following is a review of the main pump/generator architectures that affect response times and flexible power control.

4.6.3.1 Fixed-Speed Generators

Traditionally, most hydroelectric generators are synchronous machines in which their speed of rotation is directly proportional to the relatively fixed ac grid frequency depicted in Figure 4.10.

Since the speed of a turbine is mostly proportional to the flow of the water through it, there are limits to the range of the generator's output power while still maintaining a reasonable efficiency. Furthermore, the rated efficiency is specified for a specific set of conditions and power levels. The hill-curve in Figure 4.11 is a representative efficiency versus output trend that shows the peak efficiency occurring at 100% of a facility's rated output and falling to as low as 60% at low outputs. Although the magnitudes of this chart vary from facility to facility, the relationship is typical for this type of generation technology.

When the PHS facility is operating in generating mode (i.e., dropping water from the upper reservoir into the lower), it can adjust the amount of output power by mechanically restricting the flow of water into the penstocks, within certain limits. Typically, a fixed-speed generating unit can be adjusted between 30% and 100% of its rated power output; however, at 30% of its rated output, the facility is operating between 60% to 70% efficiency, dumping a lot more water per watt than at its optimal operating point.

To follow a fast-changing AGC signal,[5] a PHS facility can change its output power levels relatively quickly by adjusting the gate-valves

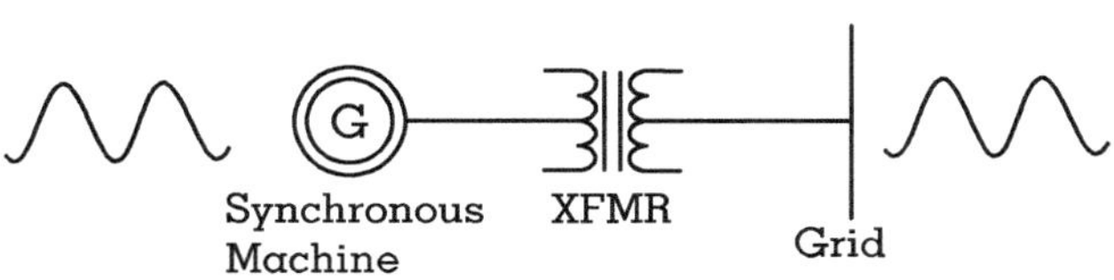

Figure 4.10 Fixed-speed, synchronous machine connected to the grid.

5. Frequency regulation signal, Chapter 2, Section 2.1.2.2.

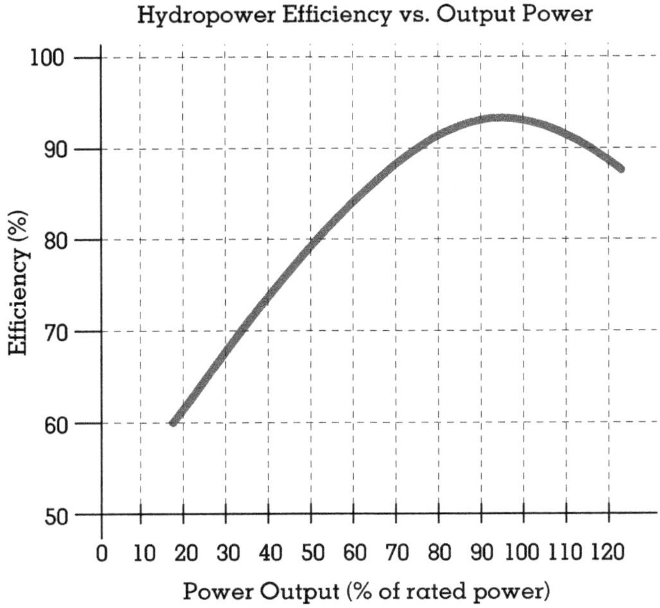

Figure 4.11 Typical shape of an efficiency vs output power curve for a hydroelectric facility (the hill-curve).

or other mechanical mechanisms that restrict the water flow through its penstocks, but only in generating mode. Unfortunately, generating modes do not last forever since the reservoir contains a finite amount of water. At some point, the generating mode will end and the upper reservoir needs to be filled again, typically on a daily basis.

When switching from one mode to the other, a fixed-speed PHS facility must close its penstocks, drain them of water, switch the motor-generator offline, wait for it to stop spinning, reverse the phase order of the grid connection, wait for it to spin up in the opposite direction, and, then finally, open the penstocks back up, all while orchestrating the sequence of water flows, gate operations, and machine controls. This entire process could take several minutes.

When fixed-speed PHS facilities are operating in pumping mode, the power absorbed from the grid is fixed. Although theoretically, one could restrict the water flowing in the penstocks while pumping, this will simply increase the pumping power as the pump struggles to push against the constriction. Therefore, fixed-speed PHS facilities pump water at a constant power set by the rating of the motor-generator and the hydraulic characteristics of the facility (i.e., head, impeller design, penstock length and width).

If the PHS facility takes several minutes to switch modes, it will not be able to follow a fast-changing frequency regulation signal very closely. Before it gets a chance to switch its mode of operation, the signal may change two or three times in polarity, not to mention vary its magnitude continuously. When the AGC is calling for pumping at different power levels, the conventional fixed-speed PHS can only put out either full power or increments of the number of pumps it has on-site. For example, if the facility has three 400-MW pumps, it can provide 400, 800, or 1200 MW of pumping power, but nothing in between. When the AGC is calling for generation, the conventional fixed-speed PHS facility will oblige only after the amount of power exceeds 20% to 40% of its generators' rating. However, when it is operating at points between 20% and 70% of its generators' power rating, it will dump much more water downstream per MWh of delivered energy than when it is operating more efficiently at full load.

Despite these limitations, PHS and hydroelectric generating facilities, in general, are still considered to be more capable of frequency regulation than large coal and nuclear power plants.

4.6.3.2 Variable-Speed PHS

When PHS facilities were first built and used, they were called upon to charge once at night and discharge through the daytime peak hours. With more renewable energy in the mix, they are called upon to reverse charge and discharge directions thousands of times a day. With the use of renewable energy ramping up in recent years, and as new revenue opportunities evolve in restructured energy markets, the value of fast-acting flexible energy storage is increasing, motivating PHS facilities to upgrade their machines to meet these new demands and opportunities.

In the past 20 years, several enhancements were made to many PHS facilities to make them more versatile. Variable-speed motor-generators and ternary-pumped hydro storage are two advanced systems that allow their operators to adjust the input and output power according to the grid's instantaneous needs.

Doubly Fed Induction Machine

A doubly fed induction machine (DFIM) can vary its speed with respect to the grid's frequency, thus allowing another dimension of control of both generation and pumping operations. Normal induction machines have a squirrel-cage rotor in which currents flow in proportion to the relative difference between the grid frequency and the spinning rotor; the higher these currents, the greater the torque on the rotor. In motoring mode, the torque causes the rotor to spin to nearly the speed of the grid's frequency. There is always a bit of slip between the rotor and the grid's frequency, which causes a finite amount of power to be dissipated in the rotor's windings. In a DFIM, the squirrel cage of the rotor is replaced by a series of windings around the rotor, which are powered by a controllable ac power source as shown in Figure 4.12.

By controlling the amount of power going through the rotor's windings, the slip, and therefore the speed of the machine, can be controlled in either generating or pumping mode.

In pumping mode, the DFIM's speed affects how much water flows uphill, which affects the electrical power coming in from the grid. In generating mode, both the speed of the generator and the flow of water through the turbine can be adjusted to vary the electrical power from the generator. With this upgrade, PHS facilities can perform variable power grid services with greater efficiencies. In 1990, the Japanese PHS plant in Ohkawachi [11] was the first to convert its motor-generators to DFIM, soon followed by others in Japan and around the world.

Although DFIM-based PHS facilities have a wider operating power range in both pumping and generating modes, it still takes nearly as long

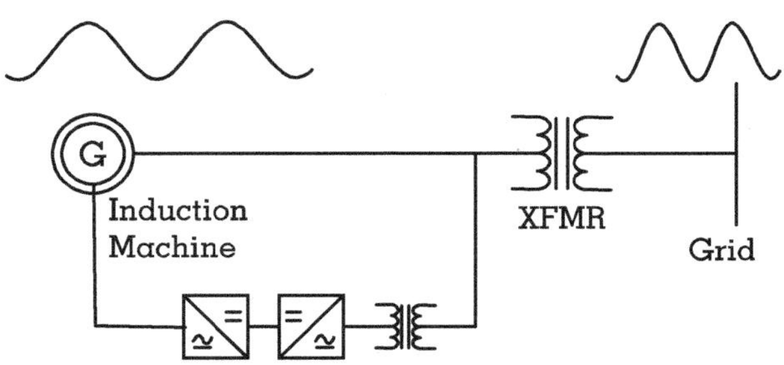

Figure 4.12 Doubly fed induction machine connected to the grid.

as before to transition between the pumping and generating modes. The process of emptying the penstocks, switching the direction of the electric machines, and starting up in reverse is still the same as for a fixed-speed PHS facility.

Convertor-Fed Synchronous Machines

Another type of variable-speed motor-generator control, referred to as converter-fed synchronous machines (CFSM), uses the same synchronous motors that are used in a fixed-speed PHS facility, but powers them with a double-conversion, back-to-back set of power converters as shown in Figure 4.13.

The first converter creates a high-voltage dc from the grid's fixed frequency voltage. The second converter then converts the dc to a variable frequency ac voltage used to power the synchronous motor-generator. The generator and pumping power are processed twice as it flows through both convertor stages. The synchronous motor-generator's speed can be completely and accurately controlled by varying the voltage amplitude and frequency of the second power converter. This type of ac/ac power conversion is used in many motor controllers, from large industrial processes to electric trains and modern wind turbines. When it was first introduced for use in hydroelectricity, it was thought that the largest system to which this could be applied was 100 MW; however, advances in power electronics are making it possible to apply to even larger systems [12]. As of this year in 2022, GE and ABB offer full double-conversion units for up to 100 MVA of power, with the words: "higher

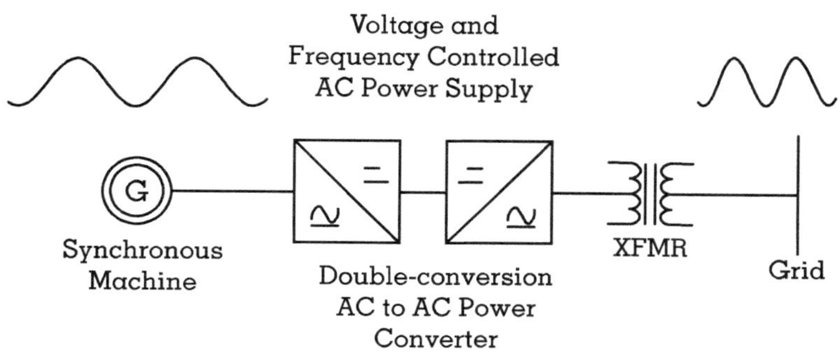

Figure 4.13 Converter-fed synchronous machine connected to the grid.

ratings available upon request" leaving the door open for higher power retrofits in the future [13, 14].

The advantage of this type is that it can transition from pump to generating modes more quickly than fixed-speed and DFIM systems. The controller simply slows the speed down to zero and changes the direction of the motor-generator to change modes. In fact, in some systems, the deceleration of the turbine is instantaneously caused by an influx of power from the grid into the electric machine. Manufacturers specify that a full switchover in direction can happen between 60 and 90 seconds.

Ternary

A mechanical way of controlling power levels is the ternary PHS with hydraulic bypass (see Figure 4.14). In this configuration, if only the turbine valves are open, water will flow from the upper to lower reservoir through the turbines that power the motor-generator which generates electricity. The turbine valves can be adjusted to vary the power output from the generator. In this mode, the clutch is open to prevent the pump from over-pressurizing its loop. If only the pump valves are open, the motor-generator will power the pump to move water from the lower to upper reservoir. While in pumping mode, if the turbine valves are par-

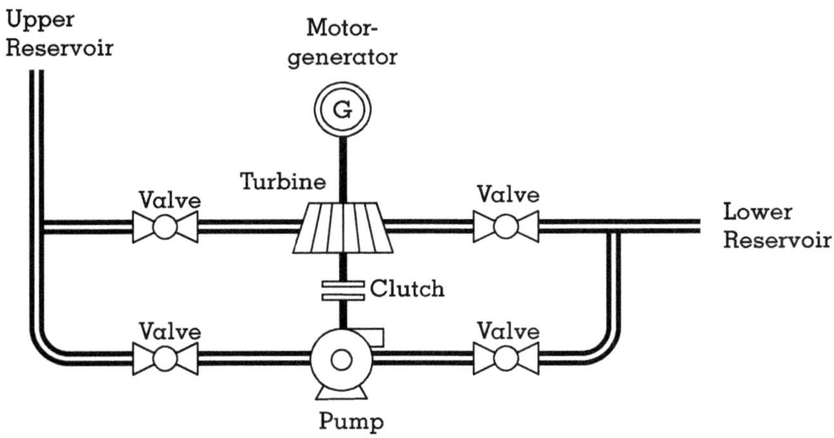

Figure 4.14 Water flow diagram of a ternary PHS with hydraulic bypass.

tially open, some water from the pump circuit will flow through the turbine and generate some electricity. This, in effect, reduces the pumping power coming from the grid, thereby offering a means to vary the charging power from 0% to 100% of the rated power of the pumps.

Besides having full control over the range of input and output power, ternary PHS can switch between pumping and generating modes in less than 30 seconds because the rotating masses never need to change direction. It is simply a matter of engaging or disengaging the clutch and opening and/or closing valves in the right sequence, so that the rediverted water can quickly change the power flow of the motor-generator.

While all pumped hydro storage facilities can perform bulk energy shifting from one part of a day to another, having the ability to quickly adjust the power flow and direction is key to being able to contribute to the short-duration power markets for grid stability, such as frequency regulation, spinning reserve, ramp management, renewables integration, and load shedding. Fixed-speed systems can only contribute to these services during generating mode, but variable-speed systems can contribute during both pumping and generating mode, making them much more valuable in this regard. Ternary PHS with hydraulic bypass are faster and more flexible but may be less efficient at partial generating power since they are still subject to limitations of their turbine speeds. Table 4.1 is a summary of the performance of the four main types of PHS technologies.

4.7 Summary of Pumped Hydro Storage

As of 2022, 95% of all electrical energy storage is performed through pumped hydro, constructed around the world since the late 1800s. They have a long service life, as proven by the continuous operation of facilities that were built several decades ago. Their round-trip efficiencies are continually improving with new plant enhancements and technologies. Updated facilities operate at around 80% round-trip efficiency at the grid point of connection. Modern plants equipped with variable-speed motor-generator controls are flexible enough to perform ancillary grid services, in addition to providing bulk energy services for which they were originally intended.

Table 4.1 Summary Performance of Different Types of PHS Technologies

Generating Mode				
Function	**Fixed-Speed**	**DFIM**	**CFSM**	**Ternary**
Power range	30% to 100%	20% to 100%	0% to 100%	0% to 100%
Start-up time (s)	75–90	75–85	65–70	65
Change to pump mode (s)	240–420	240–420	65–70	25
Short-term power markets	Yes	Yes	Yes	Yes
Voltage support	Yes	Yes	Yes	Yes
Energy shifting	Yes	Yes	Yes	Yes
Pumping Mode				
Function	**Fixed-Speed**	**DFIM**	**CFSM**	**Ternary**
Power range	100%	60% to 125%*	0% to 100%	0% to 100%
Start-up time (s)	160–340	160–230	65–70	80
Change to generator mode (s)	90–190	240–420	65–70	25
Short-term power markets	No	Yes	Yes	Yes
Voltage support	Yes	Yes	Yes	Yes
Energy shifting	Yes	Yes	Yes	Yes

* This range is dependent on the pump design.

On the other hand, PHS can have significant disadvantages as well. The construction of these huge facilities can be highly disruptive to the local environment and can take decades to complete, including planning, permitting, construction, and start-up. Pumping billions of gallons of water in and out of natural ecosystems can cause safety, ecological, and environmental problems, and their operations are subject to seasonal, and even yearly, meteorological conditions. Finally, finding a location where there is enough height and volume to store large amounts of potential energy, and where the environmental impacts are acceptably benign, is probably the biggest hurdle to this form of energy storage.

Despite these challenges, there are many proposed projects at various levels of review, permitting, and construction throughout the world, especially in China, with its goals of achieving 120 GW of PHS by 2030. In the United States, several projects are also under review, some of which are seeking to repurpose former surface and subterranean mines as water reservoirs. Table 4.2 is a summary of the general characteristics of PHS.

Table 4.2 PHS Summary Characteristics

Specification Parameter	Value	Notes
Energy density, volumetric	2.2 Wh per m^3 (vol.) per m (height)	220 Wh / m^3 for 100m height
Energy density, intrinsic	0.0022 Wh/kg per m (height)	—
Cycle life	Infinite	No degradation from cycles
Calendar life	N/A	Some reservoirs may suffer from silt accumulation over time
Round-trip efficiency	70% to 80%	Modern facilities are 80%
Self-discharge	94–370 m^3 / km^2 evaporation (Single-digit SOC loss per year)	Evaporation and soil leakage are site dependent
Operating temperature range	N/A	Generally, ground temperature is constant at bottom of reservoir, but might not work in places where deep freezes may be possible
Failure modes	Penstock rupture, machine failure, silt accumulation, turbine wear, valve seizure	Biggest past problems were surface cavitation on turbine vanes and fins
Maintenance requirements	Debris removal, bearing replacement, turbine replacements	—
Representative manufacturers (of equipment)	General Electric (U.S.) ABB (Switzerland) Andritz (Austria) Dongfeng Electric (China) Harbin Power Equipment (China) Power Machines (Russia) RainPower (Norway) Shanghai Electric (China) Siemens (Germany) Voith (Germany)	www.ge.com/renewableenergy/hydro-power/large-hydropower-solutions/hydro-turbines

References

[1] Mey, A., "Utility-Scale Batteries and Pumped Storage Return about 80% of the Electricity They Store," February 21, 2021, https://www.eia.gov/todayinenergy/detail.php?id=46756.

[2] ASME, "Rocky River Pumped Storage Hydroelectric Station," *The American Society of Mechanical Engineers*, September 13,1980.

[3] Meyer, K., "This Energy Storage is Tough on Connecticut River," March 17, 2018, https://commonwealthmagazine.org/opinion/this-energy-storage-is-tough-on-connecticut-river/.

[4] Dingley, A., "Wikimedia Commons," https://creativecommons.org/licenses/by/2.5.

[5] Grande Dixence SA, "Grande Dixence, Renewable Energy in the Heart of the Alps," www.grande-dixence.ch.

[6] TVA, "The Mountain Marvel," https://www.tva.com/about-tva/our-history/built-for-the-people/the-mountaintop-marvel.

[7] TVA, "Raccoon Mountain," https://www.tva.com/Energy/Our-Power-System/Hydroelectric/Raccoon-Mountain.

[8] US Dept of Energy, "Global Energy Storage Database," Available: http://www.energystorageexchange.org/.

[9] Koronowski, R., "The Inside Story of the Wolrd's Biggest 'Battery' and the Future of Renewable Energy," August 27, 2013, https://thinkprogress.org/the-inside-story-of-the-worlds-biggest-battery-and-the-future-of-renewable-energy-8984e81283c/.

[10] Shahan, Z., "Largest Pumped Hydro Facility in World Turns On in China," January 04, 2022, https://cleantechnica.com/2022/01/04/largest-pumped-hydro-facility-in-world-turns-on-in-china/.

[11] Kuwabara, T., A. Shibuya, and H. Furuta, "Design and Dynamic Response Characteristics of 400 MW Adjustable Speed Pumped Storage Unit for Ohkawachi Power Station," in *IEEE Transactions on Energy Conversion* Vol. 11, No. 2, 1996.

[12] Aubert, S., et al., "Variable-speed Pumped Storage with Converter-Fed Synchronous Machines (CFSM)—A High Value in Grids with Large Penetration of Wind and Solar Generation," in *HYDRO 2014*, Lake Como, Italy, 2014.

[13] GE.com, "MM7 Modular Multi-level Converter," https://www.gepowerconversion.com/product-solutions/medium-voltage-drives/mm7.

[14] ABB.com, "Variable-speed drive for converter fed synchronous machine," https://new.abb.com/power-converters-inverters/energy-storage-grid-stabilization/converters-for-pumped-storage-plants/pcs-8000-variable-speed-converter/variable-speed-drive-for-converter-fed-synchronous-machine.

Appendix 4A

Table 4A.1 List of the Largest PHS Stations in the World [8,10]

Title	Commissioned	Type	Power (GW)	Energy (GWh)	Location
Fengning PHS	1/4/2022	N.S.	3.6	N.S.	N.S., Hebei Province, China
Bath County	12/1/1985	C.L.	3.0	24.0	Warm Springs, Virginia, U.S.
Huizhou	6/15/2011	O.L.	2.4	N.S.	Huizhou, Guangdong, China
Guangzhou	6/1/2000	O.L.	2.4	20.0	Guangzhou, Guangdong, China
Dniester	1/1/1988	O.L.	2.3	N.S.	Sokyriany,Chernivtsi Oblast, Ukraine
La Muela	10/14/2013	O.L.	2.0	24.0	Cortes-La Muela, Valancia, Spain
Okutataragi	6/1/1998	O.L.	1.9	15.5	Asago, Hygo, Japan
Kannagawa	1/1/2020	O.L.	1.9	0.0	Minamiaiki, Nagano, Japan
Ludington	1/1/1973	O.L.	1.9	15.0	Ludington, Michigan, U.S.
Tianhuangping	12/25/2000	O.L.	1.8	N.S.	Anji County, Zhejiang, China
Grand Maison	1/1/1987	O.L.	1.8	N.S.	Vaujany, Isere, France
Baishan	11/1/2005	O.L.	1.8	N.S.	Huadian, Jilin, China
Jixi	12/1/2010	O.L.	1.8	N.S.	Fuling, Anhui, China
Dinorwig	1/1/1984	C.L.	1.7	8.6	Dinorwig, Wales, United Kingdom
Srisailam	1/1/1981	O.L.	1.7	N.S.	Srisailam, Andhra Pradesh, India
Raccoon Mountain	1/1/1978	O.L.	1.6	36.3	Chattanooga, Tennessee, U.S.
Mingtan	1/1/1994	O.L.	1.6	N.S.	Sun Moon Lake, Nantou, Taiwan
Tumutelectric 3	1/1/1973	O.L.	1.5	N.S.	Talbingo, New South Wales, Australia
Okumino	11/1/1995	O.L.	1.5	N.S.	Nagano, Nagano, Japan
Liyang	5/1/2011	O.L.	1.5	N.S.	Tianmu Lake, Jiangsu, China
Xianju	1/1/2015	O.L.	1.5	N.S.	Xianju, Zhejiang, China
Sardar Sarovar	6/20/2006	O.L.	1.5	N.S.	Navagam, Gujarat, India
Nakanosawa	5/1/1980	O.L.	1.3	N.S.	Omachi, Nagano, Japan
Qingyuan	7/9/2009	O.L.	1.3	11.5	Qingyuan, Guangdong, China
Castaic	1/1/1973	O.L.	1.2	12.5	Pyramid Lake, California, U.S.
Tongbai	1/1/2006	O.L.	1.2	N.S.	Hangzhou, Zhejiang, China
Hohhot	11/20/2014	O.L.	1.2	N.S.	Hohhot, Inner Mongolia, China

Table 4A.1 (continued)

Title	Commissioned	Type	Power (GW)	Energy (GWh)	Location
Helms Project	6/30/1984	O.L.	1.2	N.S.	Fresno County, California, U.S.
Okuyoshino	1/1/1980	O.L.	1.2	14.7	Kazeya, Nara, Japan
Tanbara	1/1/1982	O.L.	1.2	N.S.	Tone-gun, Gunma, Japan
Tamahara	7/1/1986	O.L.	1.2	N.S.	Numata, Gunma, Japan
Matanogawa	4/1/1996	O.L.	1.2	N.S.	Kofu, Hino District, Tottori, Japan
Zagorsk	11/5/2003	O.L.	1.2	N.S.	Sergiev Posad, Moscow Oblast, Russia
Xilongchi	10/16/2009	O.L.	1.2	N.S.	Taiyuan, Shanxi, China
Bailianhe	11/21/2009	O.L.	1.2	N.S.	Huanggang, Hubei, China
Heimifeng	10/1/2010	O.L.	1.2	N.S.	Changsha, Hunan, China
Omarugawa	7/1/2011	O.L.	1.2	N.S.	Miyazaki, Miyazaki, Japan
Baoquan	12/15/2011	O.L.	1.2	N.S.	Huixian, Xinxiang, China
Pushihe	9/29/2012	O.L.	1.2	N.S.	Dandong, Liaoning, China
Xianyou	12/20/2013	O.L.	1.2	N.S.	Xianyou, Fujian, China
Hongping	1/1/2015	O.L.	1.2	N.S.	Hongping, Jiangxi, China
Chiotas (Entracque)	1/1/1981	O.L.	1.2	N.S.	Entracque, Cuneo, Italy
Coo-Trois-Pontselectric	1/1/1979	O.L.	1.2	5.8	Trois-Ponts, Liege, Belgium
Blenheim-Gilboa Project	7/1/1973	O.L.	1.2	17.4	North Blenheim, New York, U.S.
Aldeadávila	1/1/1986	P.B.	1.1	N.S.	Duero, Salamanca, Spain
Shin Toyone	1/1/1972	O.L.	1.1	N.S.	Toyone, Aichi, Japan
Northfield Mountain	1/1/1972	O.L.	1.1	8.5	Northfield, Massachusetts, U.S.
Vianden	4/17/1964	O.L.	1.1	N.S.	Vianden, Diekrich, Luxembourg
Rocky Mountain	1/1/1995	O.L.	1.1	N.S.	Floyd County, Georgia, U.S.
Muddy Run	1/1/1966	O.L.	1.1	N.S.	West Holtwood, Pennsylvania, U.S.
Bad Creek	1/1/1991	O.L.	1.1	25.6	Oconee County, South Carolina, U.S.
Goldisthal	10/20/2004	O.L.	1.1	8.5	Thuringian Slate Mountains, Thuringia, Germany
Markersbach	1/1/1979	O.L.	1.1	4.2	Markersbach, Saxony, Germany
Imaichi	7/1/1988	O.L.	1.1	7.5	Nikko, Tochigi, Japan
Siah Bishe	1/1/2013	O.L.	1.0	N.S.	Siah Bisheh, Mazandaran, Iran

Table 4A.1 (continued)

Title	Commissioned	Type	Power (GW)	Energy (GWh)	Location
Upper Cisokan	1/1/2018	O.L.	1.0	N.S.	Bandung, West Java, Indonesia
Aguayo II	1/1/2014	O.L.	1.0	N.S.	Bárcena de Pie de Concha, Cantabria, Spain
Minghu	8/1/1985	O.L.	1.0	N.S.	Shuili, Nantou, Taiwan
Roncovalgrande (Lago Delio)	1/1/1973	O.L.	1.0	0.0	Maccagno, Varese, Italy
Drakensberg Scheme	5/1/1981	O.L.	1.0	10.0	Jagersrust, KwaZulu-Natal, South Africa
Okikuyotsu (Okukiyotsu)	1/1/1982	O.L.	1.0	N.S.	Niigata Prefecture, Honsh, Japan
Edolo	1/1/1985	O.L.	1.0	N.S.	Edolo, Brescia, Italy
Domenico Cimarosa (Presenzano)	1/1/1991	O.L.	1.0	N.S.	Presenzano, Caserta, Italy
Yangyang	9/12/2006	O.L.	1.0	N.S.	Yangyang, Gangwon, Korea, South
Tai'an	12/20/2006	O.L.	1.0	N.S.	Tai'an, Shandong, China
Yixing	12/1/2008	O.L.	1.0	N.S.	Yixing, Jiangsu, China
Zhanghewan	2/1/2009	O.L.	1.0	N.S.	Jingxing, Hebei Province, China
Lago Bianco	12/13/2010	O.L.	1.0	N.S.	Poschiavo, Grisons, Switzerland
Xiangshuijian	8/14/2011	O.L.	1.0	N.S.	Wuhu City, Anhui, China
Linthal	1/1/2015	O.L.	1.0	N.S.	Linthal, Glarus, Switzerland
Tehri	12/31/2016	O.L.	1.0	N.S.	Tehri, Uttarakhand, India

O.L. = open-loop; C.L. = closed-loop; P.B. = pump-back; N.S. = not specified.

CHAPTER 5

Thermodynamic Energy Storage

This chapter discusses energy storage that involves the principles of thermodynamics (i.e., the transfer of heat). CAES and thermal energy storage are the two main subcategories of this chapter. In the former case, heat is an integral part of the process but does not always contribute to the storage of energy, while in the latter case, heat is directly contributing to the energy storage process.

5.1 CAES

CAES uses electrical and/or mechanical power to compress air into a cavern and, when needed, releases the air back into the environment through a turbine to generate electricity. Although it seems like a simple mechanical concept, the real-world application of this technology is a complex series of electrical, mechanical, environmental, and thermodynamic processes.

5.1.1 Compressing Air

Air compressors gather air molecules from an environment of ambient temperature and pressure and force them into a smaller volume. This mechanical work adds both energy per molecule and mass to a constrained volume. Put another way, mechanical compression of a gas increases both its heat and internal energy for a given volume. Consequently, the temperature of the compressed gas rises, and the gas molecules exert more force on the constraining enclosure walls.

Utility-scale CAES systems compress the air into massive underground caverns (Figure 5.1) to as much as 70 to 100 bar. This is 30 to 40 times what is ordinarily put into a car tire. If this compression were done without removing the heat from the compressed air, the compressed air temperature could be in the range of 650°C to 800°C, which could damage the cavern structure and the compressing equipment. Additionally, the high temperature further raises the output pressure, which increases the work required to add more air into storage. As a result, most operating and planned CAES systems remove the heat from the compressed air flow before storing it.

When the compressed air is later released from the cavern to spin a turbine and generate electricity, the air temperature falls dramatically as it decompresses. The falling temperature further reduces the pressure available at the turbine, making it difficult to extract the energy stored in the compressed air. External heat is required to raise the air temperature so that it can effectively drive the turbines. This heat comes from either stored thermal energy or via burning fossil fuels.

To explain the practical implementation of compressed air energy storage in real-world plants, it will help to first review the thermodynamic processes in ideal scenarios. When compressing a volume of gas, either one of two idealized paths can be taken. One path is adiabatic,

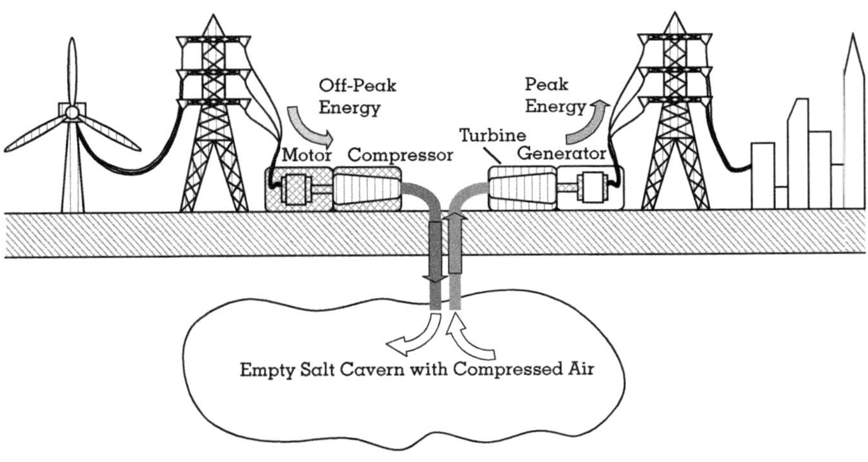

Figure 5.1 Simplified schematic of the components of a compressed air energy storage system.

meaning no heat is removed from the gas while it is being compressed. A similar process happens when an engine cylinder is rapidly compressed by a piston during the compression stroke (Figure 5.2). The compression stroke happens so fast that there is essentially no heat lost during the stroke.

The other path is isothermal, meaning that the gas temperature remains constant. Such a process can happen if a cylinder is slowly compressed, allowing the heat of the compression to be absorbed by the cylinder walls and dissipated to the environment (Figure 5.3). The physical work applied to the volume is lost to the environment, and the compressed gas is left with the same internal energy as before it was compressed, but in a more compact volume. In this process, the energy per volume is increased, but the energy per molar content remains the same.

Figure 5.4 charts the pressure versus volume for both idealized cycles compressing air. Practical values in physical machines will differ due to nonideal transfers of energy, but the curves shown are nominally representative. The path from point (1) to (2) represents an adiabatic process whose values are determined by:

$$P = P_1 \left(\frac{V_1}{V} \right)^{\gamma} \tag{5.1}$$

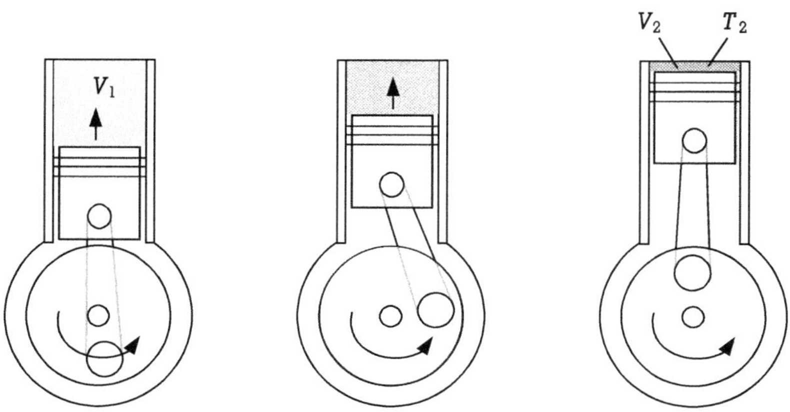

Figure 5.2 Rapid adiabatic compression in an engine cylinder.

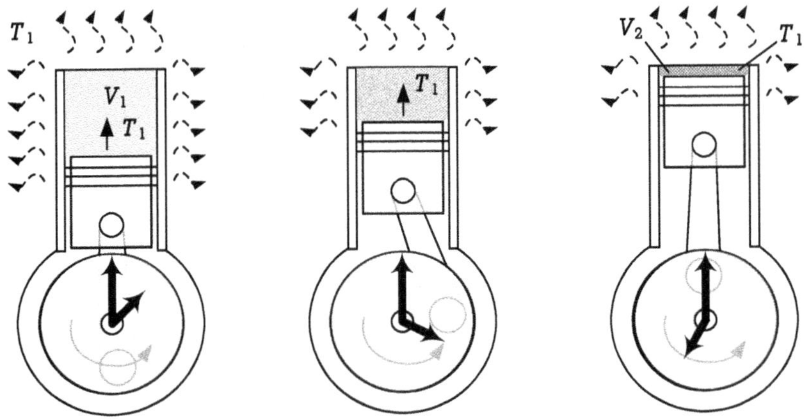

Figure 5.3 Slow isothermal compression.

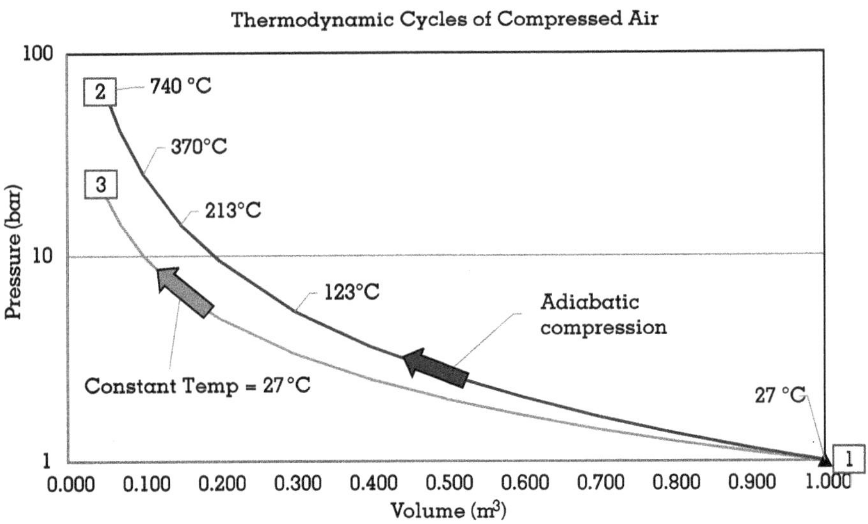

Figure 5.4 Idealized thermodynamic cycles of compressing air.

where γ, a thermodynamic property of air, is approximately 1.4.

Point (1) in Figure 5.4 corresponds to a cubic meter of air at 27°C and 1 bar. If it is compressed adiabatically, it will follow the curve between points (1) and (2) and, when it is compressed to 5% of its original volume, its pressure will be 66 bar (960 PSI). Based on the temperature of an adiabatic process,

$$T = T_1\left(\frac{P}{P_1}\right)^{\frac{\gamma-1}{\gamma}} \tag{5.2}$$

the temperature of the compressed air will reach over 700°C at this point. The compressed volume contains more molecules per unit volume and additional heat energy, and the work, W, required to perform this stroke is:

$$W = \int_{V_1}^{V_2} PdV = \frac{P_1V_1}{1-\gamma}\left[\left(\frac{V_2}{V_1}\right)^{1-\gamma} - 1\right] \text{ or } \frac{P_1V_1}{1-\gamma}\left[\left(\frac{P_2}{P_1}\right)^{\frac{\gamma-1}{\gamma}} - 1\right] \tag{5.3}$$

The mechanical energy required to adiabatically compress a 1m³ volume starting at 1 bar to 5% of its original volume is 580 kJ. Graphically, this is the area below the curve between points (1) and (2) in Figure 5.4.

Alternatively, if the cylinder were slowly compressed, allowing the heat of the compression to be absorbed and dissipated through the cylinder walls, the pressure would follow the path between points (1) and (3) in Figure 5.4 and could be calculated by:

$$P = P_1\left(\frac{V_1}{V}\right) \tag{5.4}$$

When the air gets compressed to 5% of its original volume, its pressure would be 20 bar, requiring an input mechanical energy of:

$$W = \int_{V_1}^{V_2} PdV = P_1V_1 \ln\left(\frac{V_1}{V_2}\right) \text{ or } P_1V_1 \ln\left(\frac{P_2}{P_1}\right) \tag{5.5}$$

The mechanical energy required to isothermally compress a 1m³ volume starting at 1 bar to 5% of its original volume is 300 kJ. Graphically, this is the area below the curve between points (1) and (3) in Figure 5.4. The 300 kJ is lost as heat energy to the environment and is the minimum amount of energy required to achieve 95% compression.

In a pure adiabatic compression cycle, the input work is much higher because the retained heat elevates the air temperature, which increases its pressure, which in turn pushes back on the compression stroke. This work is the maximum amount of energy required to achieve 95% compression. To reduce the amount of input work required, practical implementations of CAES shed the heat of compression either continuously or in stages, such as in a multistaged, intercooled compressor, so that the area under the pV curve is as close as possible to that of under an isothermal curve.

In one approach to a multistaged compressor, air enters the first compressor at ambient pressure and temperature, gets quickly and adiabatically compressed to a higher pressure, and exits at an elevated temperature, estimated by (5.2). An intercooler removes the added heat and lowers the temperature to somewhere near ambient. This air then enters the next compressor and again gets rapidly and adiabatically compressed to a higher pressure and temperature. The heat of this compression stage is mostly removed by another intercooler and the air enters another compressor. At the very end of the compressor train, the exiting air is cooled and sent into the cavern, where the last bit of heat is eventually absorbed by the chamber walls. This process is schematically shown in Figure 5.5.

Figure 5.6 shows the stepped pressure values versus volume. As the pressure ramps up adiabatically from point (1), at the end of the first compression stage, it drops to near the isothermal compression path and repeats this pattern twice thereafter. The work required to bring the air to point (3) is larger than that for a pure isothermal cycle, but not as much as for a pure adiabatic cycle. The extra work required is quantified by the area between the multistage compressor path and the isothermal path.

Whether the pV path is adiabatic, isothermal, or multistaged adiabatic, the work put into compressing the air, is converted to heat that is either saved in the compressed air, saved in a separate thermal storage, or dissipated to the environment.

5.1.2 Decompressing Air

Like other mechanical energy storage systems, such as in applying a force to compress a spring or elevate a mass, there is no added energy to

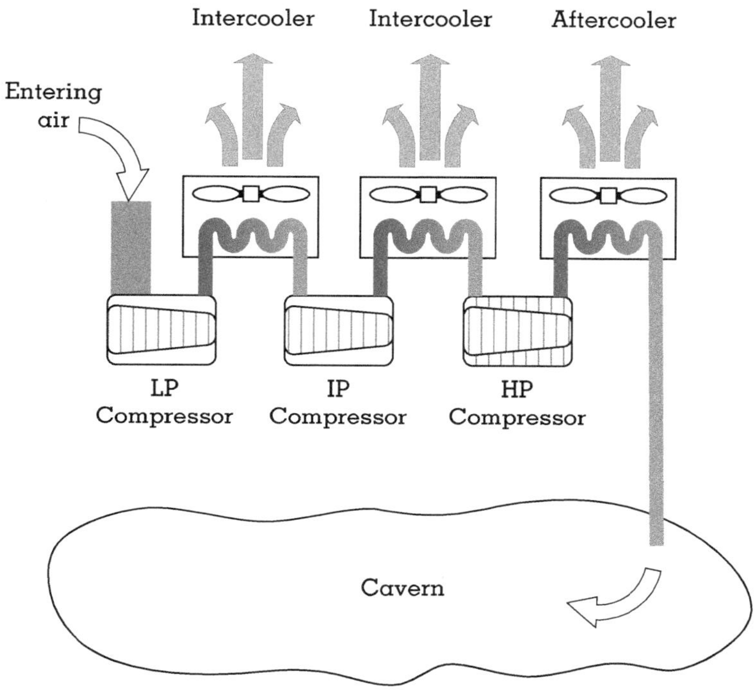

Figure 5.5 Conceptual schematic of multistaged air compressors with intercoolers between them.

the mass; rather, there is increased potential energy of the system in the possibility of having the system exert a force over a distance. Gas compressed to a smaller volume exerts a force on its enclosure, and when the gas is allowed to expand again, the potential energy is released to the environment, performing work against the moving barriers. Specifically, to retrieve the stored energy in the compressed air, it is released into the atmosphere through a turbine. If the air quickly decompresses, it starts at point (3) in Figure 5.7 and follows an adiabatic descent toward a very low pressure and temperature.

It is evident from the plot that a pure adiabatic decompression results in a dramatic reduction in pressure and temperature. The area below the curve between points (3) and (4) represents an unacceptable amount of recovered energy, not to mention extremely cold temperatures that could damage the turbines and other equipment. If heat were

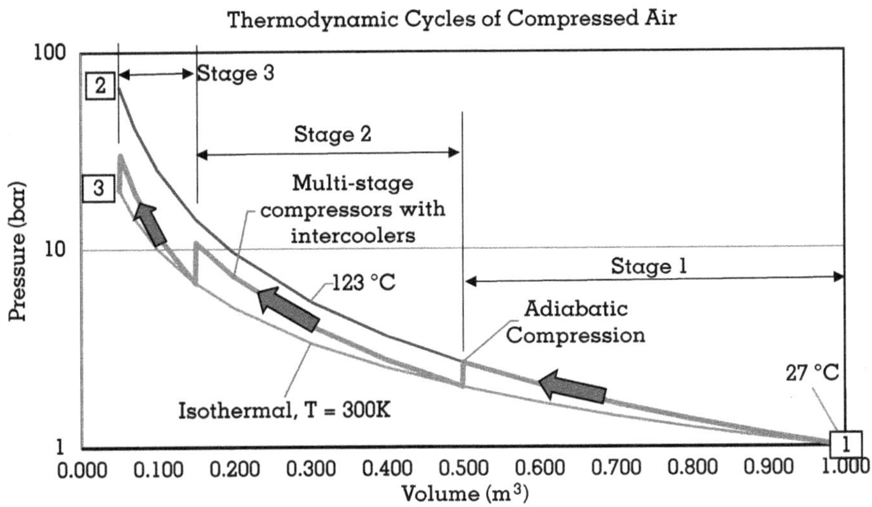

Figure 5.6 Pressure vs volume of multistaged air compression.

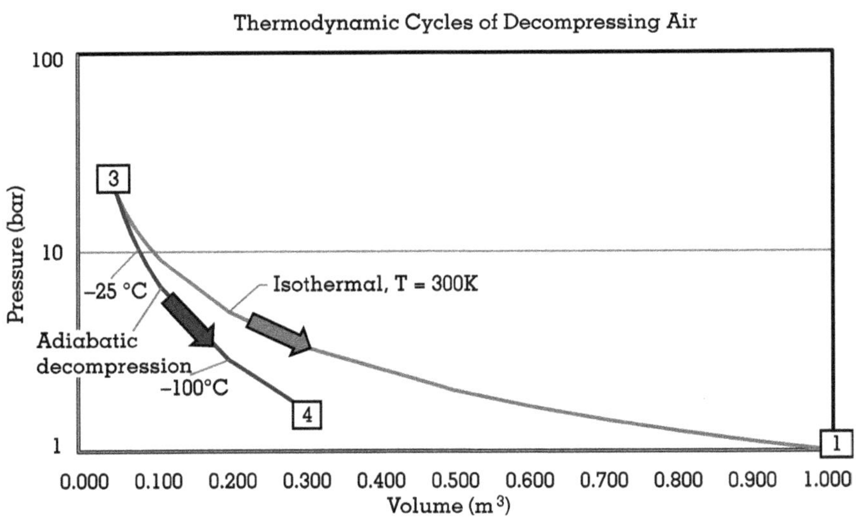

Figure 5.7 Decompressing air adiabatically and isothermally.

added to the decompressing air to maintain its original temperature, it could follow the isothermal curve from point (3) to (1). The area under this curve is the work that the decompressing air could potentially perform in a purely frictionless world. The only way that the same amount of energy can be retrieved from the compressed air that was originally used to compress the air is to retrieve the heat that was created during the compression cycle. In fact, if more heat is added to the air than originally taken out, the path would follow an even higher pV path and yield even more mechanical output energy.

Although these curves represent ideal conditions, they demonstrate the underlying principles at work during practical implementations of CAES. For one, rapid, unabated air compression generates enough heat to raise the air temperature to levels capable of damaging equipment and the storage medium. Therefore, most practical CAES systems remove this heat of compression during the charging process. Second, discharging the pressurized gas lowers the pressures and temperatures of the expanding gas precipitously, reducing the effectiveness of the exiting air and possibly damaging the generating equipment. Practical CAES plants add heat back into the exiting airstream to convert its potential energy into useful output power. Successful CAES plants effectively manage not only their air flows and pressurizations, but also heat flows inherent to their compression cycles.

5.1.3 Examples of CAES Plants

The two largest utility-scale CAES plants have been operating for several decades. The first one, in Huntorf Germany, was commissioned in 1978 and, after a couple of initial problems were ironed out, it has been operating with 90% availability [1]. According to the plant's ratings, it takes 60 MW for 12 hours to charge and can output 290 MW for up to three hours continuously; then, if necessary, provide an exponentially declining power for several hours afterward. In this plant (Figure 5.8), a motor-generator powered by the grid spins a set of compressors, which draw air from the outside and compress it to a storage pressure between 43 and 70 bar. After each compressor stage, a cooler extracts the heat of compression from the compressed air and rejects it to the environment. Almost 30% savings in energy can be achieved with a three-stage compressor versus a one-stage compressor [2].

This brings the compressed air temperature down to about 40°C before it enters the storage cavern. When electricity is needed on the

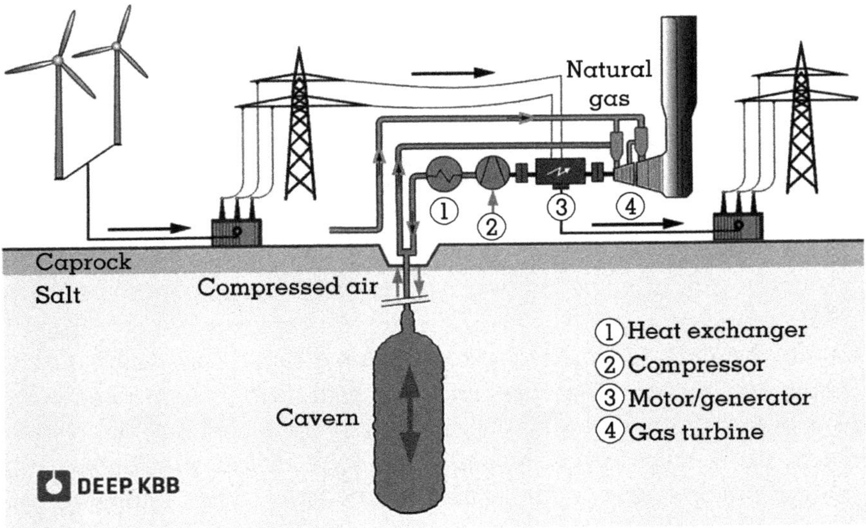

Figure 5.8 Simplified diagram of the Huntorf CAES plant. (© DEEP.KBB GmbH.)

grid, the compressed air is released into a multistage gas turbine, where it is combined with natural gas and combusted. The process of using fuel to supplement the extraction of stored energy is called a diabatic cycle. Normally a gas-powered turbine requires two-thirds of its mechanical output to compress the incoming air used in combustion. In the Huntorf plant, using precompressed air triples the output mechanical power for a given amount of combusted fuel [1]. As the plant is delivering 290 MW of power, 190 MW comes from the compressed gas while 100 MW comes from the combustion of fuel. Together, shedding the heat of compression and burning natural gas contribute to round-trip efficiencies on the order of 42% [3] for this plant.

The second largest CAES plant, installed in 1991 in McIntosh, Alabama, is similar to the Huntorf plant but uses waste heat from natural gas combustion to further warm the decompressing air to recuperate its energy content (Figure 5.9). The operating efficiency, including burning fuel, is improved to around 54% round trip [3]. This facility can provide a steady 110 MW for 26 hours when coupled with fossil-fuel burning.

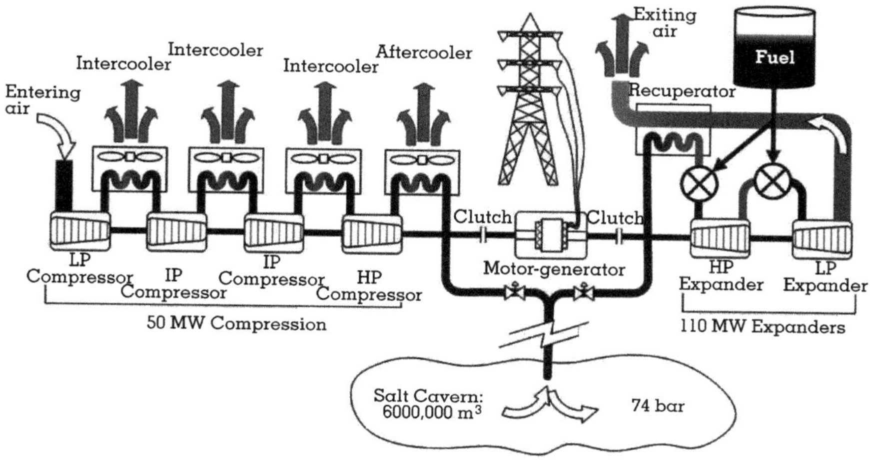

Figure 5.9 Simplified diagram of the McIntosh, Alabama CAES plant [4].

5.1.4 Siting CAES

It takes an enormous volume to store energy in the form of compressed air for utility purposes. The Huntorf facility has two caverns totaling 310,000m^3, while the McIntosh facility has a single 600,000m^3 cavern. These caverns are hundreds of meters below the surface, embedded within large salt deposits. To create them, workers drilled a hole into the salt layer and pumped fresh water into the cavern to dissolve a portion of the salt around the pipe. After the water became saturated with salt, the briny mixture was pumped out and discharged into the sea. This process was executed gradually, so as not to affect the environment, and was continuously repeated for two years until the caverns were large enough. This process is not unique to this industry since caverns of this type are used to store natural gas to help maintain its availability from season to season. The two caverns created for Huntorf are about 250m tall, 20m to 40m wide, and 650m below the surface. Figure 5.10 shows the irregular shapes and sizes of these caverns from sonar and laser scans.

However, there are a few complications with using salt caverns. The decompressing air from the cavern contains salt, which is highly corrosive to the pipes, turbines, and other equipment handling the air. Therefore, special consideration must be taken in selecting materials for plant construction. Huntorf learned this the hard way when their initial

Figure 5.10 Shape and relative size of the two Huntorf caverns created by laser and sonar scans. (© DEEP.KBB GmbH.)

equipment corroded after just a few years, leading to a major reconstruction with upgraded materials (ceramic coatings, stainless steel, etc.). With these upgraded components, the plant continued to operate for another 40 years and is still in operation as of 2022.

Other types of sites have been considered for CAES storage, including abandoned oil and natural gas wells, natural gaps in the earth's crust layers, abandoned mines [5], porous rock, aquifers, and even aboveground tank farms. For subterranean options, many factors, such as available size, geologic porosity, depth, structural integrity, oxygen reactions, and means of brine disposal have to be considered [4]. According to one study done by EPRI, 80% of the United States contains favorable geologic conditions for CAES caverns (Figure 5.11).

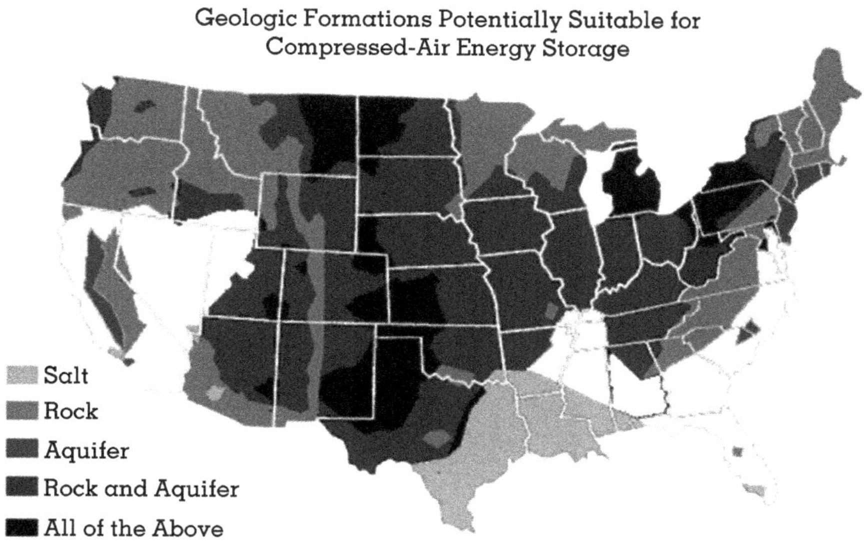

Figure 5.11 Map of the United States, showing potentially favorable geologic conditions for subterranean CAES sites. (From: [6].)

Finally, as is the case in all large energy storage facilities, proximity to electrical transmission lines, infrastructure, and load centers is an important consideration.

5.1.5 Advanced CAES Technologies

Several approaches have been considered to overcome the thermodynamic challenges of compressing air. The diabatic cycles of the Huntorf and McIntosh facilities reject most of the heat of compression while charging and use fuel to add heat back into the process when discharging, and thus have very low round-trip energy efficiencies. Alternatively, some developers have proposed using a pure adiabatic cycle, which stores both the heat and the compressed air in an enclosure that can withstand extremely high temperatures and pressures. For the most part though, combinations of both adiabatic and isothermal processes are the focus of several active research projects and commercial ventures in this field.

5.1.5.1 Advanced Adiabatic

In a pure adiabatic cycle, heat is retained in the compressed air as it is compressed, elevating its temperature, pressure, and the amount of input mechanical energy. Another form of an adiabatic cycle is to move the heat from the gas to thermal storage while charging, and then move the heat from storage back to the gas when discharging (see Figure 5.12). This reduces the amount of energy required to store the gas and increases the energy that can be recovered from it, without relying on an external heat source. The cost of doing this is in the added heat storage containers and exchangers, and the amount of energy required to move the heat during a cycle. In some designs, it may be as simple as passing the compressed gas through a heat exchanger associated with a large thermal mass. In other designs, the heat may be moved by additional heat-pump equipment that transfers the heat to and from a molten salt heat reservoir.

Saving and reusing the heat of compression eliminates the need to artificially heat the decompressing air with fuel combustion, and thus increases the total round-trip efficiency. Storelectric [7] is one such compa-

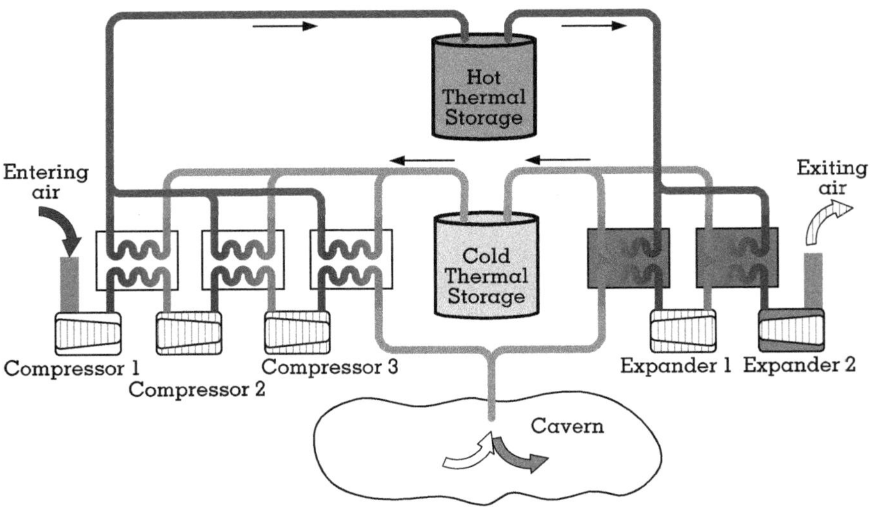

Figure 5.12 Simplified diagram of a CAES equipped with TES for advanced adiabatic operation.

ny using this technology and claims between 62% and 70% total system round-trip efficiency.

5.1.5.2 Isobaric

Another variant of the CAES process is one in which the pressure in the cavern is maintained by the hydrostatic pressure of a reservoir situated on the surface (Figure 5.13). As the cavern is alternatively filled and emptied during the charge and discharge cycles, the water from the surface reservoir maintains near-constant cavern pressure through a connecting pipe. The pressure in the storage cavern depends only on the average depth of the cavern with respect to the surface reservoir. As in other CAES types, there are similar heat flows, but the difference is that the pressure remains constant, while the volume of the storage space changes during a cycle. The area under the pV curve is simply the pressure times the fully charged, air-filled volume; therefore, the amount of energy that can be stored for a given volume is much higher than for a constant-volume

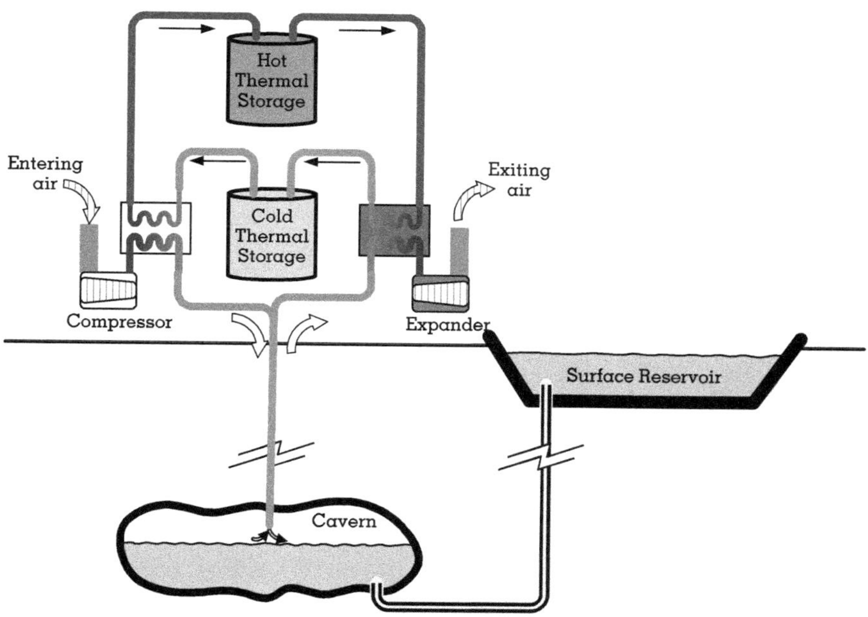

Figure 5.13 Simplified schematic of a CAES using thermal storage and isobaric compression.

CAES. Another advantage is that the output power throughout the charge and discharge cycles can be constant, which greatly increases the versatility of the storage asset.

Hydrostor [8] is a company using this technology on a couple of its upcoming projects that are expected to be online between 2025 and 2027. Three large projects in Australia and California will total over a GW, each having 8 hours of storage (Figure 5.14). In their proposed projects, the heat of compression will be saved for later discharges. A pool of water at the surface, connected via pipes to the air cavern below, will provide the constant pressure in the cavern. In addition to the energy stored in the pressurized gas, there will be some energy stored in the water that is raised and lowered between the surface reservoir and cavern in each cycle. Hydrostor claims between 60% and 75% round-trip efficiency for their storage [9].

5.1.5.3 Others

Alternative cycles of CAES proposed by others utilize different variations of the thermal and combustion cycles.

In a *steam generator combined cycle* plant, waste heat from the combustion process converts water to steam, which drives another

Figure 5.14 Rendering of one of Hydrostor's upcoming projects using thermal storage and isobaric compression. (Source: Hydrostor Inc. Reprinted with permission.)

turbine that generates additional electrical power from the plant. This is considered a short-term, power-boost operation because it cannot sustain the continuous transfer of heat to drive the steam generator.

In a *steam injection cycle*, rejected heat from the combustion is used to heat water to steam that is then mixed with the decompressing air before it reaches the turbines. This does two things: it adds heat to the cooling air stream, and it adds mass to the flow into the turbines. Both of these increase the discharging power of the turbines.

The *humidification cycle* is similar to the steam injection cycle in that water is added to the expanding air flow, but in this case it is not added as steam but as a water mist. The added water in the air flow greatly increases the power of the turbine simply due to the added mass of the water on the turbine blades.

There are many more types and combinations of those listed, such as:

- Combined humidification and adiabatic heat recuperators;
- Adiabatic systems that generate steam to create electrical power during the compression stage;
- Various adiabatic processes using auxiliary compression/decompression chilling systems to transfer the heat;
- Those that use heat absorption chillers powered by the intense heat of compression.

5.1.6 Controlling CAES

Since compressors and turbines are two different machines in a CAES plant, they can be sized independently. By having two separate machines, the plant can be rated to charge at a different power than it discharges. For example, Huntorf is rated to charge at 60 MW but discharges at 290 MW, while the Mcintosh plant is rated to charge at 50 MW and discharge at 110 MW. Furthermore, the compressor and turbines could be connected to two independent electric machines to further increase the plant's versatility.

Despite their complexities, CAES facilities can start generating power within a few minutes of their start-up command. They can ramp to 100% power in 3 to 5 minutes, and they are also capable of starting up without the presence of grid power and thus can help to start up other power plants after a widespread blackout. The Huntorf plant in Germany has been used as a black-start resource on multiple occasions in its 40+ year service.

CAES plants can easily vary their output power by adjusting the rate of fuel burned and compressed air released. Some of the more complex CAES systems that use steam generators and other intertwined thermodynamic processes may not be as flexible, but they will be more efficient at full power output. This variability in output power and their ability to rapidly switch directions makes them a valuable resource for frequency regulation and response.

Another advantage, as Mark Howitt from Storelectric points out, is that CAES systems have large inherent inertia built into their grid-connected power generators in the form of synchronous machines. These machines have large inertia built into them because they often are linked on the same axle to several spinning turbines in multistaged expanders. This inherent inertia helps stabilize the grids from momentary losses of large generating plants or transmission assets.

5.1.7 Summary of Compressed Air Energy Storage

In the early 2000s, CAES was the second-largest energy storage provider, second only to PHS, and it achieved that with only two large storage facilities. These plants have been providing peak energy services for many decades. Despite not being the most efficient energy storage plants, they provide a useful load shifting and stabilization benefit to the grids they serve. Since then, batteries have overtaken CAES partly because of their higher efficiency, higher power, site-ability, and other advantages. However, the promises of adiabatic and isobaric improvements on the conventional compressed air energy storage, having efficiencies approaching 70%, may encourage a different look at CAES for the future. CAES, in conjunction with TES, can not only improve this technology's efficiency, but also can eliminate the need for additional fuel consumption, simplify its installation, and increase its reliability.

Despite no major additions of CAES since 1991, there is a recent resurgence of interest in it as the need for energy storage increases, as is evidenced by several California, Canada, and Australia projects being developed as of 2022. The advantages of a very long service life and relatively small surface footprint weigh positively against its low efficiency and challenging siting requirements. Table 5.1 summarizes the general characteristics of compressed air energy storage.

5.2 Thermal Energy Storage

Thermal energy storage (TES) is an extremely broad field that includes the storage and use of heat for many purposes. Although TES does not make the news every day as the next millennial, technological innovation to go public, it is the most common form of energy storage in people's daily

Table 5.1 Compressed Air Energy Storage Characteristics

Specification Parameter	Value	Notes
Energy density, constant volume	~1 kWh / m^3 cavern (constant volume storage*, †)	*Very little surface land required for subterranean cavern storage †Assume diabatic or a-adiabatic with TES
Energy density, constant pressure	4 Wh/m^3 (vol.)/m (depth) e.g., 4 kWh/m^3 for a 1,000m deep cavern	Includes energy in displaced water
Cycle life	Infinite	No degradation from cycles
Calendar life	N/A	—
Round-trip efficiency	43% to 72%	Depending on type of cycle
Self-discharge	Negligible	Site dependent
Operating temperature range	N/A	Generally, ground temperature is constant in cavern
Failure modes	Parts corrosion, machine wear, geologic shift	Biggest past problems were pipe and turbine corrosion from salty air
Maintenance requirements	Periodic mechanical machine maintenance	—
Representative manufacturers (of equipment)	Hydrostor Storelectric MAN Energy Solutions (Germany) Siemens AG General Electric AlaCAES Magnum Development Apex CAES Hydrodynamics Group LLC	hydrostor.ca www.storelectric.com www.man-es.com www.siemens-energy.com www.ge.com alacaes.com magnumdev.com www.apexcaes.com www.hydrodynamic-group.com

lives. Almost every home and business has one: a hot water tank. Hot water storage enables the availability of domestic hot water whenever it is needed and minimizes the size requirements of the hot-water heater.

Heat can be collected from the sun, the electric grid, a fossil-fuel-powered heat source, compressing air, or industrial heat waste. The heat travels via fluid, gas, steam, or even fine sand particles [10] to a storage tank where, depending on how well it is insulated, it can stay for hours or weeks. Heat can either be used directly for comfort and industrial processes, or indirectly to generate electricity by spinning a turbine, powering a Stirling Engine, driving a heat pump, or radiating a thermo-photovoltaic panel (generic process shown in Figure 5.15).

This chapter will focus on TES as it applies to large grid systems, and not systems that provide heat (or refrigeration) to homes, businesses, and communities directly. That said, domestic heating and cooling thermal energy storage have opportunities to improve the grid's power balance. Using the inherent thermal storage in building structures or adding additional bespoke thermal mass, electric-powered comfort thermal management systems can be coordinated with the grid's condition to help improve its stability in short periods or even in diurnal cycles.

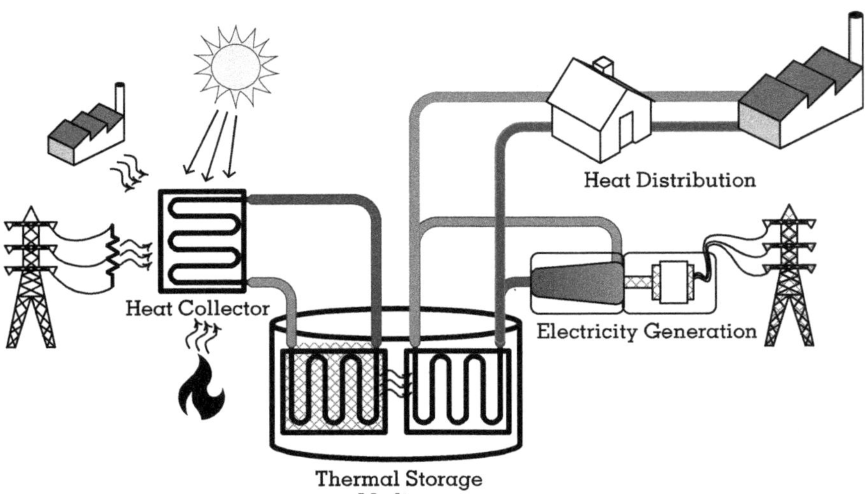

Figure 5.15 Schematic of a generic thermal storage system.

There are three main categories of thermal energy storage: sensible, latent, and thermochemical. Sensible storage is where added heat simply increases the temperature of a storage medium. The amount of energy that can be stored in a storage medium is directly proportional to its specific heat capacity, c_p, and its temperature rise. Its energy can be felt or sensed. Latent thermal energy storage uses the heat buried in the phase changes of a material. When a material transitions from a solid to a liquid, or a liquid to a gas, a large amount of heat is absorbed by the material. The phase change happens without the material changing its temperature, so it has some beneficial aspects in its applications, aside from the large amount of energy it can absorb for a given volume. Finally, thermochemical energy storage systems use reversible chemical reactions that are exothermic in one direction and endothermic in the other. The advantage of this type of storage is that to maintain the state of charge, one simply has to keep the reactants separate from each other. To extract the energy, one simply combines the reactants, and heat is emitted (or absorbed) and put to use.

5.2.1 Sensible Thermal Energy

The effectiveness of sensible thermal energy storage relies on the ability of the storage medium to contain heat. The specific heat capacity, or c_p, is the ratio of the heat that can be stored with respect to its mass. Two other factors are the range of working temperature it can withstand, ΔT, and its material density, ρ. Since the specific heat capacity can vary over the temperature range under consideration, the heat is calculated by integrating it with respect to temperature, over the whole temperature range, T_{low} to T_{hi}.

$$Q_{sensible}/V = \int_{T_{low}}^{T_{hi}} \rho \cdot c_p \cdot dT \tag{5.6}$$

5.2.1.1 Solid Media Thermal Storage

Concrete, bricks, table salt, and rocks have thermal capacities ranging between 0.9 and 1.1 kJ per kg and can be heated to several hundred degrees Celsius. Iron and steel alloys have even lower thermal capacities, but since they are extremely dense, they have some of the highest thermal capacity per unit volume. The theoretical volumetric energy densities are calculated based on the temperature limits of the materials up to their melting points, which does not consider the practicalities of enclosure designs and available heat sources. Table 5.2 lists the thermal energy properties of a few of the commonly considered solid storage media.

5.2.1.2 Liquid Media Thermal Storage

Liquids, in general, have lower energy density per unit volume than solid TES materials, but they are useful for transporting heat from one place to another, something that cannot easily be done with solid media.[1] Some liquids are also used to store heat such as molten salts and silicon. Table 5.3 lists the thermal energy properties of a few of the commonly considered liquid storage media for large-scale purposes (water is included as a reference).

Table 5.2 Solid Media Thermal Energy Properties

Solid Materials	**Specific Heat Range (25°C to T_{hi}) (kJ/kg/ K)**	**Density* (kg/m³)**	**Upper Temperature T_{hi} (°C)**	**Sensible Heat† $Q_{sensible}/V$ (MJ/m³)**
Concrete	0.90–1.10 [11]	2,400	500	897
NaCl	0.86–1.05 [12]	2,160	1,074 (mp)‡	1,178
Sand (contains phase transition) (c_p data used is granite)	0.77–1.16 [13]	1,600 (packed sand)	1,000	1,586
Granite (contains phase transition)	0.77–1.16 [13]	2,650	1,000	2,322
Magnesia fire bricks (data used is for MgO)	0.92–1.33 [12]	3,580	1,200	5,126
Stainless steel 316	0.50–0.68 [14]	8,000	1,427 (mp)	5,420
Iron (contains phase transition)	0.45–0.65 [12]	7,874	1,200	6,088
Graphite	0.71–2.13 [12]	2,260	2,000	7,970
α Al_2O_3	0.77–1.34 [12]	3,930	1.800	8,437

*Density is assumed to be relatively constant.
†See Table 5A.1 for how $\rho \times c_p$ was calculated.
‡mp = melting point.

1. A notable exception is sand, which can be transported through pipes along with a hot air flow.

Table 5.3 Liquid Media Thermal Energy Properties

Liquid Materials	Specific Heat Range (T_{low} to T_{hi}) (kJ/kg/K)	Density (kg/m^3)	Temperature Range (T_{low}, T_{hi}) (°C)	Sensible Heat per Volume $Q_{sensible}/V$ (MJ/m^3)
Water	4.18* [12]	~1,000*	25, 100	314
Na-0.79K metal eutectic	0.98–0.88 [15]	867–699 [16]	25, 700	482
Heat transfer oil (Philips Heat Transfer Fluid 32)	1.90–2.88 [17]	862–688 [17]	25, 288 [17]	495
Silicone oil (Syltherm™ 800)	1.6–2.3 [18]	991–547 [18]	25, 400 [18]	540
Therminol™ VP-1	1.56–2.63 [19]	1070–694 [19]	25, 400 [19]	687
Nitrate salts (60/40 $NaNO_3/KNO_3$)	1.5[b] [20]	1905–1724 [21]	290, 560 [22]	735
Na (liquid)	1.3–1.25 [23]	921–783 [23]	98, 700	765
Silicon	0.97[b] [12]	2521–2199 [24]	1412, 2400	2258
NaOH	2.15–2.04 [12]	1783–1446 [21]	323, 1100	2603

*Assumed to be relatively constant.

5.2.1.3 Concentrated Solar Plant Thermal Energy Storage

Sensible heat storage is the most widely used form of energy storage for electrical grid purposes. For example, as described earlier in this chapter, TES is used to store the heat of compression during the charging cycle of CAES and release that heat into the expanding air flow during the discharging cycle. In addition, concentrated solar plants (CSP) are using it to provide a more continuous flow of power to the grid. Since 1984, when the first CSP plant was installed in the United States, energy storage has been increasingly coupled with CSP. In fact, the total amount of global CSP-coupled energy storage is over 22 GWh, most of it built in the years between 2010 and 2020 [25] (Figure 5.16). More are in the planning and constructions stages beyond 2021. As of 2021, there are currently over 116 CSP operational plants in 22 countries.

The average power rating of the operating CSPs as of 2020 is 50 MW, but they can range from 300 kW to 400 MW. Larger ones are in the planning stages. About half of the installed CSPs have energy storage attached to them. The average energy rating of a CSP is 7 hours, but they range from between 15 minutes to 17 hours of energy at full power.

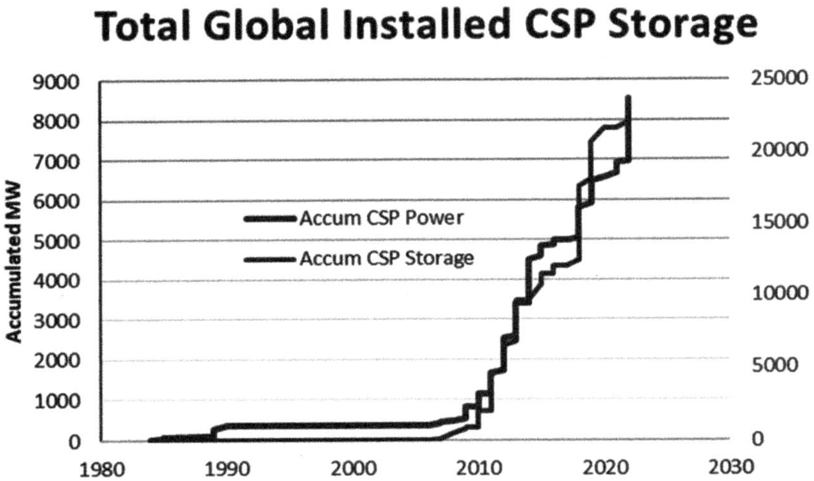

Figure 5.16 Installations of concentrated solar plant generation and storage. (Data source: [25].)

In a CSP, the sun heats the heat transfer fluid to over 400°C through a collection of mirrors that concentrate the solar radiation on a narrow focal point where the fluid is exposed (Figure 5.17). The fluid is typically a specially formulated thermal oil, molten salt, or steam (see Table 5.3). The heat in the fluid is either transferred to a storage tank for later use or directly to a steam turbine where it immediately creates electricity (Figure 5.18). Electricity in a CSP is usually generated by a spinning synchronous machine powered by a Rankine turbine that is spun by a steam loop through a heat exchange with the storage medium or heat transfer fluid.

If the heat transfer fluid is an oil, it goes through a heat exchanger to heat a bath of molten salts or other heat storage media. Salts are usually chosen for the storage medium because of their lower cost compared to oil, but they bring other challenges. First, they must be kept above their freezing temperature, or the whole plant could shut down and require that each pipe be heated and drained. Second, the salts are highly caustic and damaging to metallic containers and plumbing. The Crescent Dunes plant (Figure 5.19) was plagued with leaking molten salt issues before it eventually shut down in 2019.

Figure 5.17 Array of parabolic mirrors, part of a CSP in Hawaii. (Source: Xklaim, Wikimedia Commons.)

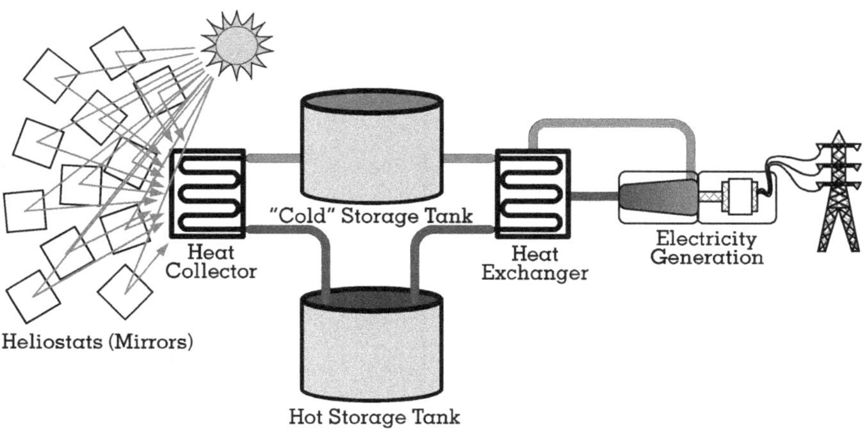

Figure 5.18 Schematic of a CSP generator and thermal energy storage plant.

Figure 5.19 Array of heliostats directing sunlight on a central tower at Crescent Dune CSP. (Source: Grikas, M., Wikimedia Commons.)

5.2.1.4 Temperature Differential Thermal Storage

Instead of adding heat to a storage unit, another way of increasing thermodynamic potential energy is to separate heat from cold. For example, a heat pump can be powered by electricity to pump heat from a tank of antifreeze liquid to a tank of molten salt (Figure 5.20). A wide difference in temperature between two storage mediums allows for more storage, higher power, and higher round-trip efficiency. The discharge cycle involves bringing the stored heat and the cold together in a heat engine, where the difference in temperatures drives a turbine and creates electricity. In 2019, Malta claimed around 60% round-trip efficiency for a full charge and discharge cycle [26]. That is pretty good considering that the absolute maximum ideal Carnot[2] one-way efficiency for their selected temperature range is in the low 70s.

2. The highest one-way efficiency that any thermodynamic process can achieve is limited to the ideal Carnot efficiency, which is $\eta_{Carnot} = \frac{T_{max} - T_{min}}{T_{max}}$ (temperatures in degree Kelvin).-

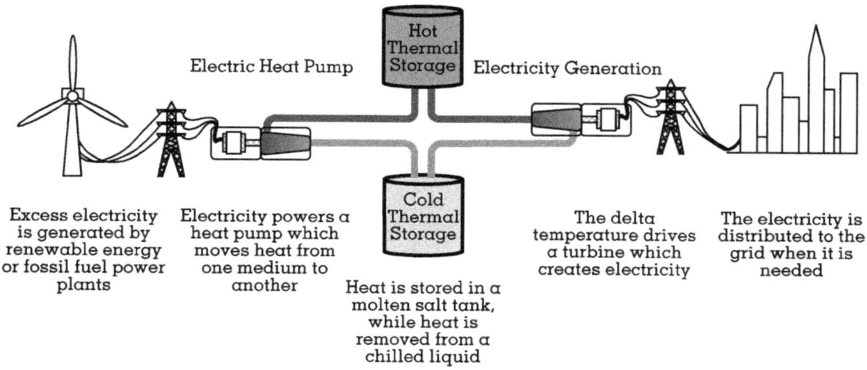

Figure 5.20 Schematic of thermal energy storage technology using separation of heat.

5.2.1.5 Creating Electricity from Heat

Most thermal-electric storage systems use a steam turbine to capture the energy of the heat to spin a generator. This is a tried-and-true technology used by virtually every fossil fuel and nuclear power plant in the world. Heat creates steam which drives a turbine that spins the generator. The efficiency of this process in a standard fossil fuel-burning plant is 35% to 40%, but there are continuously improving designs aimed at improving this metric [27]. For example, some processes use the residual waste heat to preheat the incoming water to improve the heat-to-steam process. The McIntosh CAES used this technique to reduce their fuel consumption by 25% over the Huntorf CAES, which did not use it. Using very high temperatures and expensive materials that can withstand them is another way to improve the one-way generating efficiency.

MIT is experimenting with another way to create electricity from heat. Their system uses a photoelectric junction like that of a PV cell, but tuned to convert infrared radiation, instead of visible light, to electricity. Efficiencies of around 30% or more have been cited for just the thermophotovoltaic cell [28]. Although the realizable efficiencies of this technology are still low, it does involve far fewer moving parts and mechanical complexities.

5.2.2 Latent Heat Storage

Latent heat is the heat required to change the phase of a substance from either a solid to a liquid or a liquid to a gas. In a solid material, as more heat is added, a linear rise in temperature will be observed until it reaches a plateau. At this point, the atoms or molecules break free from their intermolecular bonds and the material transitions into a liquid state, all while the temperature of the material remains relatively constant. The amount of heat required to loosen the intermolecular bonds is called the heat of fusion. For various metals and salts, this could be hundreds and even thousands of joules per kilogram. It is typically in the range of one-half or one-third of the energy required to get the material up to its melting temperature from 25°C. With aluminum, for example, the heat needed to bring it to 660°C is 1,300 kJ/kg. Once there, another 400 kJ/kg is required to melt it.

An energy storage system storing latent heat has the advantage of being able to store 1.3 to 2 times the energy in a storage medium for a given storage temperature. This reduces the insulation required, the material stresses, and a host of other technical challenges with an otherwise higher-temperature system. Silicon is a notable example, requiring almost twice the heat to melt it than to bring it to its 1,410°C melting temperature. Table 5.4 is a list of some materials considered for latent energy storage.

5.2.2.1 Integration of TES with a Fossil-Fuel Power Plant

Research into the integration of thermal energy storage into a large thermal power plant was conducted by the Argonne National Laboratory [30]. They looked at the possibility of using certain customized materials that could easily be integrated into existing plant designs. The melting temperatures of a few of the listed materials can be adjusted by mixing them with other materials. This variability can allow some flexibility when incorporating thermal energy storage into an already-existing thermal plant. For example, the alloy 70:30:10 Al:Cu:Zn has a melting temperature of 535°C (see Table 5.4). Different ratios of alloys not only can change the melting temperature but can raise the latent heat and volumetric energy density of the storage medium.

Tuning the melting temperature to match the output steam temperature of a fuel boiler allows integration of the storage with relatively

Table 5.4 Thermal Properties of Some Materials Considered for Latent Heat Storage

Storage Material	Specific Heat Range from 25°C to Melting (kJ/kg/K)	Melting Temp. (°C)	Density[c] (kg/m^3)	Sensible Heat from 25°C[d] (MJ/m^3)	Latent Heat of Fusion (kJ/kg)	Latent Heat to Melt (MJ/m^3)
Aluminum	0.90–1.18 [12]	660	2,620	1,745	397	1,040
Al Alloy (20Cu10Zn)	0.71–0.94 [29]	535	3,624	1,258	284	1,029
KBr	0.44–0.57 [12]	734	2,740	951	215	589
Table Salt (NaCl)	0.86–1.15 [12]	801	2,160	1,653	481	1,039
NaOH	1.49–2.17 [12]	323	2,130	1,125	160	341
Nitrate salts (60/40 NaNO3/KNO3) (contains phase transition)	0.90–3.42 [20]	222	1,950	593	100	195
LiF	1.61–2.39 [12]	848	2,640	4,475	1,044	2,756
Silicon	0.71–1.04 [12]	1,410	2,330	2,980	1,800	4,194

[c] Density assumed to be relatively constant.
[d] Sensible heat calculated using (5.6) and Table 5A.1.

few changes to the design of the power plant. For example, if an existing supercritical steam boiler outputs steam at a temperature of 550°C to a turbine, the TES's temperature could range between 500°C and 550°C. Selecting the 70:30:10 Al:Cu:Zn alloy with a melting point in the 535°C range [29] will provide significant latent heat storage in that medium.

The latent heat to melt the alloy at 535°C would be:

$$\begin{aligned} Q_{latent} &= \Delta H_{fus} \cdot \rho \\ Q_{latent} &= 284 \cdot 3624 = 1030\, MJ/m^3 \end{aligned} \tag{5.7}$$

If the power plant's generator turbine is 40% efficient, each cubic meter of aluminum alloy could store 114 kWh_e[3] of energy. If the target storage is four hours at a thermal output of 1.5 GW_t, the storage medium would require 21,000 m^3 of this Al alloy. Since roughly 55 m^3 fit into a fully insulated 53-ft container, this 2.4-GWh_e storage plant would require 383 × 53-ft containers. This example uses discrete container-sized storage enclosures for comparison purposes, but in a real-world plant, the volumetric efficiency could be dramatically improved if the storage medium is held in one large, insulated tank.

The advantage of this approach is that it uses an already existing power plant design that has already maximized its thermal to electrical efficiency. The only additional losses in efficiency are those related to loss of energy through heat leakage from the storage tanks. The thermal energy generated by the fossil fuel or nuclear plant is directly stored, with no appreciable transfer losses in that process. Although the underlying thermal to electrical efficiency is subject to the usual thermodynamic limitations, the energy storage part of that process is relatively seamless.

5.2.2.2 Integration of TES with a Nuclear Power Plant

TerraPower and GE-Hitachi [31] are looking at ways to capture the thermal energy of a base-load nuclear power plant before it is converted to electricity. They are using the latent heat of molten salts to capture and store the heat from a nuclear reactor in a large, insulated tank. When electricity is needed, the salt heats a steam loop which drives a turbine-generator (Figure 5.21). With enough internal storage, the power plant can provide daily load cycling while its thermal plant output can be maintained at a steady pace, thus maximizing its efficiency, and reducing thermal stresses.

5.2.2.3 Cryogenic Energy Storage

Up to this point, this chapter has described energy storage processes that raise the temperature of a storage medium, thereby increasing its energy content. Another approach is to raise the potential energy content by reducing the temperature of a storage medium. Although this seems coun-

3. The *e* subscript denotes a unit of electrical power or energy, while the *t* subscript denotes a unit of thermal power or energy resulting from a combustion process, for example, kWh_t times thermal-to-electrical efficiency = kWh_e.

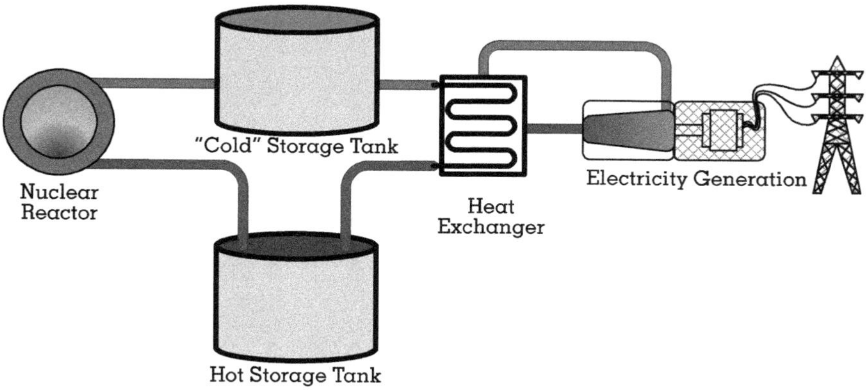

Figure 5.21 Terrapower and GE-Hitachi on the Natrium™ team are looking at ways to integrate molten salt TES into a nuclear power plant.

terintuitive, the thermodynamic potential energy stored in a material is affected by its delta temperature (and any latent heat energy associated with phase changes that occur in that temperature range), and not necessarily its absolute temperature.

In Highview's energy storage approach (Figure 5.22), called liquid air energy storage (LAES), air is cooled to the liquid state. Nitrogen and oxygen liquify at around −196°C and −183°C, respectively, and are kept at that temperature in insulated vessels for use at a later time. When electricity is needed, the super-cooled liquid is pumped to a turbine and combined with atmospheric air where it expands 700 times as it boils and turns the turbine.

The latent heat between liquid and gas of air is about 202 kJ/kg.[4] The sensible heat from (5.6) is:

$$Q_{sensible} = 1\frac{kJ}{kg \cdot K} \cdot \left[25^\circ - (-196^\circ)\right] = 221 kJ/kg$$

The theoretical total heat removed from the air to get it from 25°C to the liquid state at −196°C is 423 kJ/kg, which is a respectable 370[5] MJ/m³. Applying the maximum Carnot efficiency rule, the maximum

4. Weighted average of heats of evaporation of nitrogen and oxygen.
5. 370 MJ/m³ = 423 kJ/kg × 880 kg/m³ (density of liquid air) × 0.001 MJ/kJ.

Figure 5.22 Artist's rendition of a 50-MW / 500-MWh cryogenic thermal energy storage from Highview. (Courtesy: Highview Power. Reprinted with permission.)

one-way efficiency that can be achieved with a relative temperature difference of 25°C to −196°C is:

$$\eta_{th} < \frac{297 - 77}{298} = 74\%$$

According to Highview [32], their round-trip efficiency approaches 60%, which includes pumping, refrigeration, machine losses, and electrical losses in the process of storing and discharging energy in one cycle.

5.3 Thermochemical Storage

The idea behind thermochemical storage is that heat is used to dissociate a substance into two or more reactants, which can then be stored in separate containers. In the charged state, the reactants can be physically separated and stored for many months, or even indefinitely, unlike other forms of energy storage that leak over time. When electricity is needed, the two substances are brought together and, with some coercion, react exothermically. The heat of reaction would then be captured and used to spin a turbine or power a thermophotovoltaic converter. The input heat could come from the sun or directly from the output of a power plant. Some substances, such as certain carbonates, react and dissociate at very high temperatures, over 1,000°C. Such a high temperature, if controlled, can yield a higher thermodynamic efficiency in the process of creating electricity. Other materials, such as silica gels, exothermically adsorb water; however, since the adsorption temperature is too low to produce electricity, this technology is only considered for domestic hot water uses.

There are many challenges with this kind of storage, ranging from cyclability to cost, temperature, chemical stability, and others. As such, it is still in development and, as of this printing, has not been implemented at a commercial scale.

5.4 Summary of Thermal Energy Storage

Thermal energy storage is pervasive in the world around us. It is used in homes, businesses, industry, and power plants for hot water, heat, and even air-conditioning. Northern countries use it to capture summertime heat to supplement their winter heat requirements. Geothermal systems use it to a certain degree, using the surrounding earth as a giant thermal storage pool.

There are some trials of electrical-thermal-electrical-energy storage, where electricity is first used to charge the medium with heat (or cold), and later use that energy to generate electricity, but their round-trip efficiencies are in the 40% to 60% range, and they have not been adopted for wide-spread commercial use yet.

However, thermal energy storage for grid applications is becoming increasingly prevalent in combination with other heat sources and preexisting thermodynamic processes of generating electricity. In these cases, adding storage can extend the usefulness and increase the flexibility of thermal power plants. In the case of concentrated solar plants, thermal energy storage allows the CSP to continue to provide power after the sun has set. In the case of standard fossil fuel or nuclear power plants, thermal energy storage can allow the plant to maintain a constant fuel burn, even though the grid may be taking variable power throughout the day.

Thermal energy storage has also found a use in improving the efficiency of compressed air energy storage. By saving the otherwise lost heat of compression and using it to boost the output of the decompressing air, CAES systems no longer have to rely on burning fuel to recover stored energy. Such an approach is being developed by several companies around the world in planned installations.

Whether storing energy in compressed-air or high-temperature blocks, the direct generation of electricity is always subject to the Carnot efficiency limitation. To improve the efficiency, researchers look for

Table 5.5 Thermal Energy Storage Characteristics

Specification Parameter	Value	Notes
Energy density, constant volume	100–1,500 kWh_e per m^3	Storage medium only, not including power equipment
Cycle life	Varies by materials	Most are infinite, but some degrade with excessive heat
Calendar life	N/A	—
Round-trip efficiency	50% to 70%	Temperature difference defines max. Carnot efficiency
Self-discharge	Negligible to 10% / day	Design dependent
Operating temperature range	N/A	Operating temperatures are so much higher than ambient anywhere
Failure modes	Parts corrosion, machine wear, heat stress, freeze	Biggest past problems were pipe storage corrosion from molten salts
Maintenance requirements	Periodic mechanical machine maintenance	—
Representative manufacturers (of equipment)	Azelio Brenmiller Energy Ltd. Kraftblock GmbH EnergyNest Highview Power Terrapower Natrium General Electric	www.azelio.com bren-energy.com kraftblock.com energy-nest.com highviewpower.com www.terrapower.com www.natriumpower.com www.ge.com
CSP equipment manufacturers	Brightsource Abendgoa Siemens Acciona Sener	www.brightsourceenergy.com www.abengoasolar.com energy.siemens.com www.acciona.com www.energy.sener/solar

ways of increasing the absolute operating temperature or the temperature spread between storage media. Advances in materials science are enabling these higher temperatures and gradients, as is evidenced by the recent excursion through the sun's upper atmosphere by an earth-launched explorer. Table 5.5 summarizes the general characteristics of thermal energy storage.

References

[1] Crotogino, F., L.U. Mohmeyer, and R. Scharf, "Huntorf CAES: More than 20 Years of Successful Operation," Orlando, FL, 2001.

[2] https://www.engineeringtoolbox.com/horsepower-compressed-air-d_1363.html.

[3] http://www.eseslab.com/ESsensePages/CAES-page.

[4] Pedrick, G., "Compressed Air Energy Storage Engineering and Economic Study," New York State Electricity and Gas, Binghamton, NY, 2009.

[5] "C.A.E.S. Technology," The Hydrodynamics Group LLC, 2000, http://hydrodynamics-group.net/norton.html.

[6] Rastler, D., "Electricity Energy Storage Technology Options," EPRI, Palo Alto, CA, 2010.

[7] www.storelectric.com.

[8] www.hydrostor.ca.

[9] Wilsons, I., "Compressed Air Can Compete and Thrive in Lithium Age: Q&A," *BloombergNEF*, March 3, 2020, https://about.bnef.com/blog/compressed-air-can-compete-and-thrive-in-lithium-age-qa/.

[10] Schlipfa, D., et al., "Using Sand and Other Small Grained Materials as Heat Storage Medium in a Packed Bed HTTESS," in *SolarPACES* 2014, 2015.

[11] Zehfuß, J., et al., "Evaluation of Eurocode 2 Approaches for Thermal Conductivity," *Civil Engineering Design*, Vol. 2, 2020, pp. 58–71.

[12] Chase, M. W., "NIST-JANAF Thermochemical Tables 4th Edition," *Journal of Physical and Chemical Reference Data Monographs*, Vol. 9, American Institute of Physics, 1998.

[13] Hartlieb, P., et al., "Thermo-Physical Properties of Selected Hard Rocks and Their Relation to Microwave-Assisted Comminution," *Minerals Engineering*, Vol. 91, 2016, pp. 34–41.

[14] Kim, C. S., "Thermophysical Properties of Stainless Steels," Argonne National Laboratory, Argonne, IL, 1975.

[15] O'Donnell, W.J., P. G. Papanikolaou, and C. B. Reed, "The Thermophysical and Transport Properties of Eutectic NaK Near Room Temperature," Argonne National Laboratory, Argonne, IL, 1989.

[16] Jackson, C. B., "Liquid Metals Handbook, Sodium-NaK Supplement (TID-5277)," Washington, D.C.: Atomic Energy Commission, Bureau of Ships, 1955.

[17] Phillips 66 (R), June 2020, Phillps66Lubricants.com.

[18] Dow Chemical, "Syltherm 800 Heat Transfer Fluid," 1997, http://www.dow.com/heattrans.

[19] Eastman, "Therminol VP-1 heat transfer liquid," November 2019, Therminol.com.

[20] Costa, S. C., et. al., "Solar Salt Latent Heat Thermal Storage for a Small Solar Organic Rankine Cycle Plant," *Journal of Energy Resources Technology*, Vol. 142, 2020.

[21] Janz, G. W., "Thermodynamic and Transport Properties for Molten Salts," *Physical and Chemical Reference Data*, Vol. 17, No. 2, 1988.

[22] Bonk, A., et al., "Enhancing the Thermal Stability of Solar Salt up to 600°C in Extended Lab-Scale Experiments," in *AIP Conference Proceedings*, 2020.

[23] Fink, J. K., and L. Leibowitz, "Thermodynamic and Transport Properties of Sodium Liquid and Vapor," Argonne National Laboratory, Argonne, IL, 1995.

[24] Mukai, K., Z. and Yuan, "Measurement of the Density of Molten Silicon by a Modified Sessile Drop Method," *Materials Transactions*, Vol. 41, No. 2, 2000, pp. 323–330.

[25] National Renewable Energy Laboratory, "Concentrating Solar Power Projects," 2021, solarpaces.nrel.gov.

[26] Collins, L., "Google X Spin-off Malta could Change World, but Lags Behind Rivals," *Recharge*, September 6, 2019.

[27] "Thermal Efficiency of Rankine Cycle," https://www.nuclear-power.com/nuclear-engineering/thermodynamics/laws-of-thermodynamics/thermal-efficiency/thermal-efficiency-of-rankine-cycle/#:~:text=Most%20efficient%20and%20also%20very,reheat%20reach%20ab.

[28] Fan, D. et al., "Near-Perfect Photon Utilization in an Air-Bridge Thermophotovoltaic Cell," *Nature Magazine*, Vol. 586, 2020, pp. 237–241.

[29] Wei, G., et. al., "Thermophysical Property Measurements and Thermal Energy Storage Capacity and Analysis of Aluminum Alloys," *Solar Energy*, Vol. 137, 2016, pp. 66–72.

[30] Sienicki, J. J., et al., "High Efficiency and Low-Cost Thermal Energy Storage System," September 29, 2017, https://publications.anl.gov/anlpubs/2017/10/139393.pdf.

[31] www.terrapower.com.

[32] Collins, L., "Liquid-air storage offers cheapest route to 24-hour wind and solar," *Recharge*, July 2019.

Appendix 5A

Table 5A.1 The Temperature Dependent Models of the Product of c_p and Density

Material	Fit Temp. Range (°C)	a (x 10^{-3})	b	c (x 10^{3})	d (x 10^{7})	R^2	References
Al_2O_3	25–2,054	1.66	6.18	–3.71	7.53	0.9835	[12]
Aluminum	25–660	1.64	3.45	–4.33	26.0	0.9998	[12]
Al Alloy (20Cu10Zn)	25–535	4.88	–11.3	11.7	—	1.0000	[29]
Graphite	25–2,027	–0.444	8.43	–4.83	9.47	0.9971	[12]
Heat transfer oil (Philips heat transfer fluid 32)	16–288	0.463	3.10	–3.10	13.5	1.0000	[17]
KBr	25–734	1.01	1.03	–1.56	10.9	0.9999	[12]
LiF	25–848	2.37	8.45	–7.87	30.9	0.9990	[12]
MgO	25–1500	1.62	7.53	–6.54	19.6	0.9944	[12]
Na (liq.)	127–627	1.58	–0.929	0.283	0.414	1.0000	[23]
NaCl	25–800	1.61	1.15	–1.16	7.97	0.9999	[12]
Na-0.79K metal eutectic	25–1,000	1.07	–0.942	0.760	–2.74	1.0000	[15, 16]
NaOH (liq.)	323–1,027	4.64	–1.41	0.101	—	1.000	[12, 21]
NaOH (solid)	25–323	7.75	–35.2	82.5	–540	0.9964	[12]
Nitrate salts (liq.) (60/40 NaNO3/ KNO3)	290–560	3.44	–1.01	—	—	1.0000	[20, 21]
Therminol™ VP-1	12–400	0.872	3.30	–1.81	–14.6	0.9999	[19]
Silicon (liq.)	1412–2,400	2.97	–0.316	—	—	1.0000	[12, 24]
Silicon (solid)	25–1,412	1.14	2.25	–1.65	4.61	0.9960	[12]
Silicone oil (Syltherm™ 800)	10–400	1.53	0.663	3.18	–42.4	0.9999	[18]
Stainless Steel 316	25–1,427	3.67	1.06	—	—	1.000	[14]

CHAPTER 6

Electrochemical Energy Storage

Batteries store and deliver energy through electrochemical reactions that are enabled by the flow of electrons in an external circuit. The internal reactions can be chemical reactions, or intercalation of ions into atomic lattice structures. In every cell chemistry, ions flow inside of the battery cell, while electrons flow in a circuit outside the cell (Figure 6.1). The energy being stored and delivered is the product of the electric current and the voltage between the cell's terminals.

While discharging a lithium-ion cell, for example, lithium ions are disassociated from electrons at the negative electrode (anode), travel through the cell, and become reunited with electrons at the positive electrode (cathode). Meanwhile, the separated electrons travel through a wire from the negative terminal of the cell, do work, such as light a bulb, and return to the positive terminal of the cell. A rechargeable battery will undergo the reverse cycle while charging, but instead of the electrons doing work on a load, they are forced in the opposite direction by an electromotive force, such as a power supply. The storage process happens almost instantaneously inside the cell, and the amount of energy that can be stored in a specific volume is higher than in most other forms of energy storage. The potential to do work is caused by the difference in the electrochemical energy states of two materials that can be reversibly converted by a chemical reaction enabled by an electric current. Specific examples of how this happens, and the details of various battery types will be described later in this chapter.

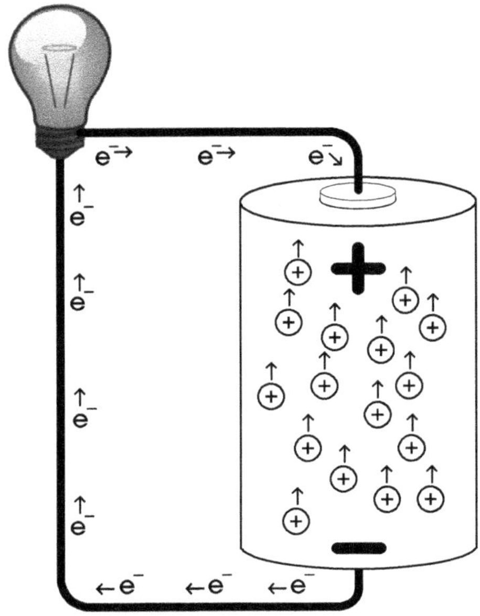

Figure 6.1 Ions flow inside an electrochemical cell while electrons flow outside the cell, shown here in a direction that powers a load.

6.1 Six Critical Components of a Battery Cell

There are six essential components of a battery that influence its performance characteristics. Although, its chemistry is its defining attribute, the successful implementation of any chemistry depends on a well-executed interplay of all six of its critical components.

> *A Note on Terminology:* A battery cell is a unit of electrochemical energy storage, having two terminals that connect to an external circuit or other cells and the rest of a storage system. A battery is a collection of cells wired in parallel or series to deliver more energy, power, voltage, and/or current than a single cell could. For example, a commonly used AA battery is not actually a battery, but a cell comprising only one electrochemical unit and providing 1.5V at its terminals. A 9V battery on the other hand, *is* a battery, having a series of six small 1.5V cells inside its rectangular enclosure. An automotive starter battery is a series of six 2V lead-acid cells in a single monobloc case.

The first critical component of an electrochemical cell are the *ions* that flow inside the cell. Lithium atoms are extremely electroactive, so it is no surprise that lithium-ion batteries have very high power and energy densities. In some cell chemistries, multiple ionic types transfer charge from one electrode to another in an ionic "bucket brigade." Finally, some ions give up more than one electron in electrochemical reactions (multivalent) resulting in more electric current for a given number of ionic transfers.

The second is the *electrolyte* through which the ions travel. Electrolytes are solids or liquids positioned between the two electrodes, which allow the transport of the ions, but not electrons. The ease by which the ions flow through the electrolyte is just one of the many factors that affect the cell's power and round-trip efficiency.

The *anode* (negative electrode) accepts electrons from the external circuit when the battery is charging. When they get there, they can be reunited with a positive ion inside the cell or cause a negative ion to split off and head to the cathode.

The *cathode* (positive electrode) accepts electrons from the external circuit when the battery is discharging. The ions simultaneously move through the electrolyte to maintain charge balance. Every transferred electron involves ions traveling to its respective electrode to maintain charge balance.

Fifth, the *separator* is a material that keeps the electrodes from touching each other internally. It is porous to absorb electrolyte, and allow ions to permeate freely, but insulating to prevent any electrons from short-circuiting between the electrodes.

Finally, the cell *case* protects the cell from the environment, keeps its internal components intact, and sometimes promotes an effective mechanical design of the battery pack. The most popular form of a battery case is a cylinder, like that of an AA cell, and is chosen for its user-friendly replaceability. Rectangular shaped, prismatic cases are used most often for lead-acid batteries and applications where physical robustness is required. Lithium-ion batteries in cell phones and other slim-line electronic devices are sealed in flexible polymer pouches chosen for their higher energy density and flexible form factors. Figure 6.2 shows schematically the relative position of each component.

In practice, a cell will be a lot more complex, containing many layers of sandwiched electrodes, separators, electrolyte, and insulators.

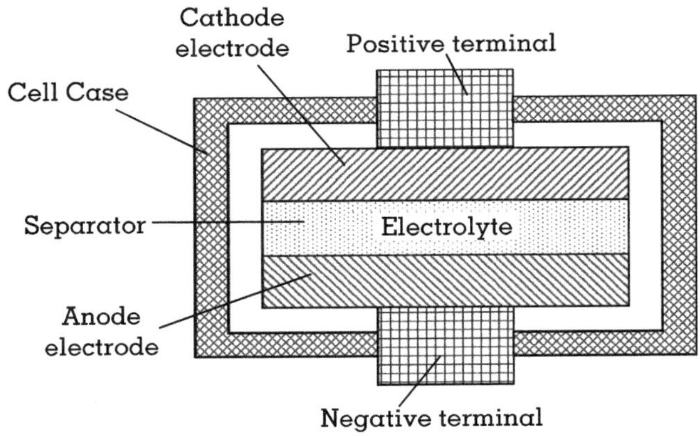

Figure 6.2 A simplified schematic of a battery cell showing its six critical components.

Large prismatic cells are made with multiple layers of plates, sandwiching layers of electrolyte-sodden separators. Originally, cells were made by dunking two mechanically constrained electrodes in a bath of electrolyte. Later, separators were added to allow the plates to be pressed close to each other for added energy density and reduced internal resistance. Cylindrical cells are made by wrapping a long sandwich of thin layers around a mandril to make a jelly-roll electrode pair.

There are many references explaining the fundamental principles of battery cells; [1] is one example, so this chapter is not going to dwell too much on those details, except suffice it to say that each of the critical components of a cell are vital to determining its characteristics, from its voltage, to its energy, to its power, and its life performance.

This is a field of ever-improving science and engineering, where the ideal energy storage capacity is always out of reach, yet tantalizingly close. Whenever the industry produces a product that makes consumer gadgets do so much more than they did yesterday, customers beg for more. Electric cars with 40 miles of range were followed by those with 80, then 150, then 400 and greater. Cell phones that were carried around in a briefcase are now thrown in jacket pockets and lost in the folds of airline seats. Dick Tracy's smartwatch would have never been possible without today's battery technology.

6.2 Battery Energy Storage Architecture

Before diving into the different battery chemistries and types, this section will summarize the architecture of a typical battery energy storage system. This presents a sense of how the cells, modules, racks, and packs go together to create large, multimegawatt systems that store gigawatt-hours of energy.

Battery cells have terminal voltages that are too low for grid energy storage purposes, so they are connected in series to yield higher, workable voltage levels (see Figure 6.3). For example, many energy storage systems operating at the megawatt level will have a nominal terminal voltage between 500V and 1,500V. A 1,000V nominal dc bus voltage will require around three hundred 3.3V cells in series. Typically, this is achieved by connecting a string of 15 to 30 modules in series, each module having 10 to 20 cells in series.

The energy in a string is correlated with the capacity of its constituent cells:

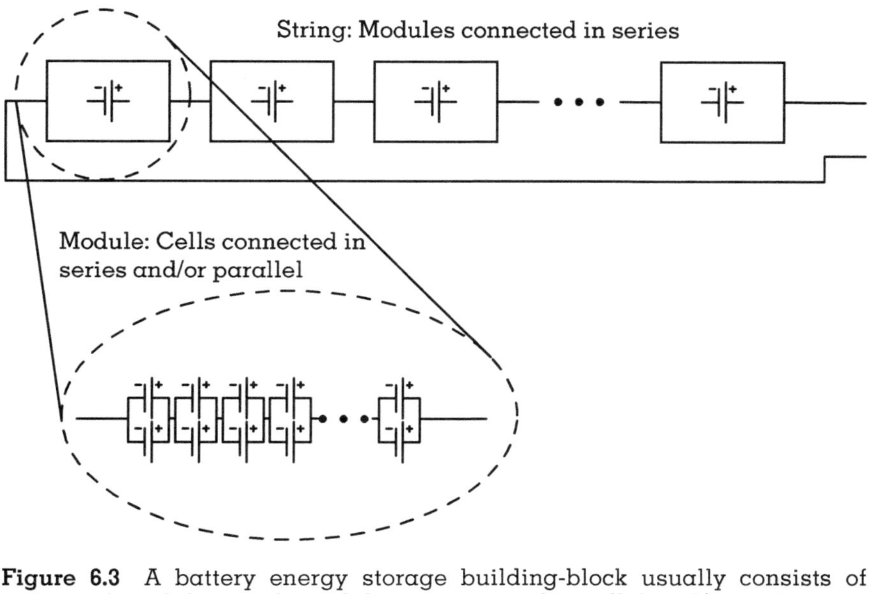

Figure 6.3 A battery energy storage building-block usually consists of strings of modules, each module consisting of parallel and/or series connected cells.

$$E = nominal\ voltage(V_{nom}) \times capacity(Ah) \tag{6.1}$$

To achieve 1 MWh of storage, a 300V string must have 3,333 Ah of capacity. If this is not practical for mechanical or electrical reasons, multiple strings of cells can be connected in parallel, for example to achieve a megawatt-hour of energy storage, ten 100-kWh strings may be connected in parallel (see Figure 6.4), each 100-kWh string having three hundred 3.3V, 100-Ah cells in series.

$$1\ \text{MWh} = 10\ \text{strings} \times 300\ \text{cells} \times 100\ \text{Ah} \times 3.3\text{V}$$

These numbers are just examples of how the general principles of scaling energy storage are calculated. The exact calculations depend on a whole host of other factors related to the cell chemistry, electrical design, system, environment, and mechanical design to name a few.

To connect a dc battery to the grid, an electronic power conversion system converts the battery's dc voltage to the grid's ac voltage (see

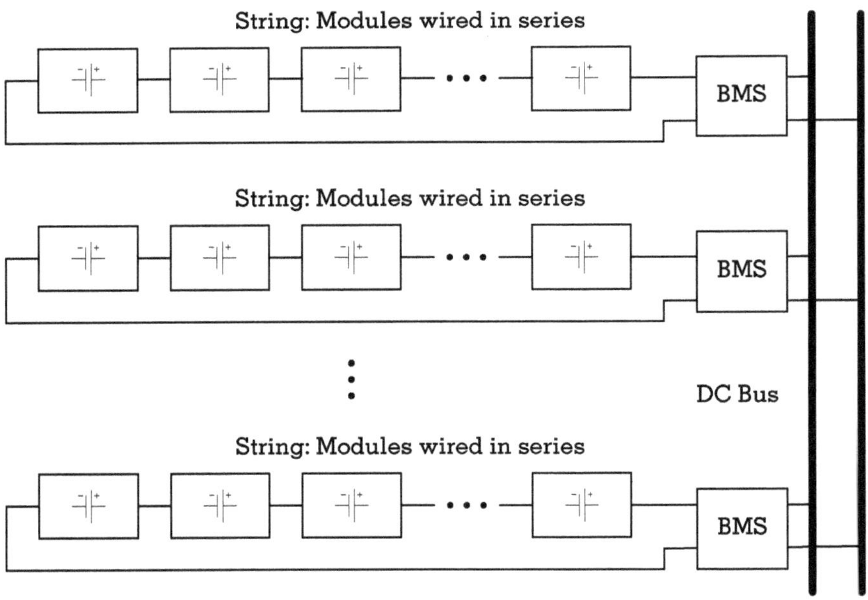

Figure 6.4 A battery energy storage system usually consists of parallel-connected strings of modules, each string protected by an independent BMS [2].

Figure 6.5). By adjusting the relative difference between its ac output and the grid voltage, electronic power conversion systems (PCS) can regulate the power flow between their inputs and outputs. In Figure 6.5, the "DC Battery" symbol represents any dc battery, whether it might be a lead-acid monobloc or a container of lithium-ion battery racks.

Bidirectional converters can force power to go both directions, between the battery and the grid. In contrast, PV PCSs are designed to transfer power in only one direction: from the PV string to the ac grid. Multiquadrant PCS can not only move real power to and from the battery, but they can adjust the phase of the ac current with respect to the grid voltage to move reactive power in and out of the grid simultaneously. This is useful to help regulate grid voltage and compensate for long transmission lines and other nonideal conditions found on a typical power grid.

Larger battery energy storage systems are usually arranged in groups of energy storage components, each independently controlled by a separate PCS (see Figure 6.6). This allows the separate groups to be managed according to their needs, current status, and/or state of health. It also provides a level of redundancy that allows the rest of the groups to continue service if any one of the groups goes down for maintenance or fails. Typically, a group of PCS such as that shown, will be interconnected to a collection bus at medium voltage (e.g., 13.8 kV), which is then connected to a distribution grid, or in larger sites, transformed to a higher voltage (e.g., 115 kV), and connected to a transmission line.

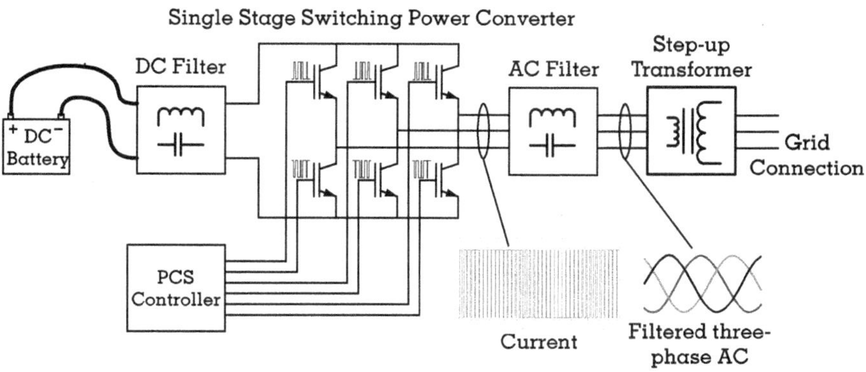

Figure 6.5 A battery energy storage system is usually connected to its loads or the grid through an electronically controlled power conversion system [2].

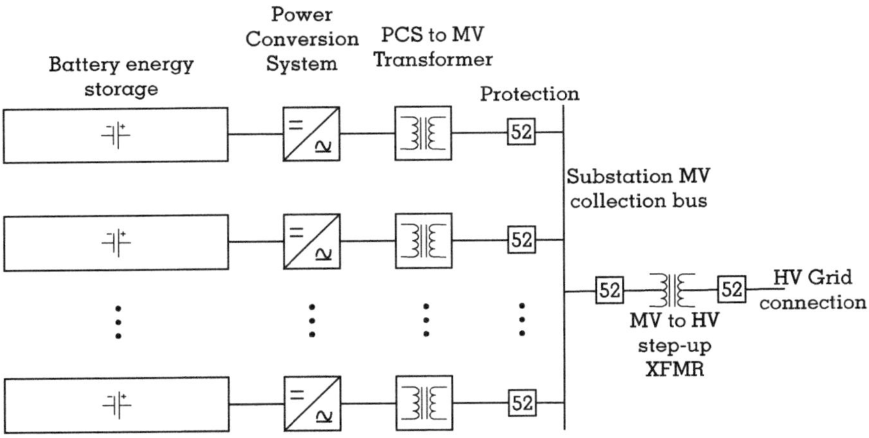

Figure 6.6 Multiple-battery energy storage systems are connected to the grid through parallel-connected power conversion systems for redundancy and flexibility [2].

Not shown, but implicitly included, are the many layers of controls and other auxiliary subsystems required for proper operation of any energy storage system. Auxiliary power is usually taken from the grid in a separately metered tap to provide power to controls, communication systems, monitoring and protection systems, and environmental control systems. Environmental control is critical to most battery systems to keep them operating in their sweet spot of performance and service life. Many times, this comprises an HVAC system that removes the heat from around the battery due to its round-trip cycle losses and dissipates it to the environment outside the enclosure.

Additionally, an ESS will have a central controller (sometimes referred to as an energy management system or EMS) that manages the general operations of the equipment and may communicate to the grid operator to update its status and receive commands. The EMS is usually responsible for receiving commands from the grid operator and doling them out to the power converters according to their abilities and/or their connected batteries' conditions. Successful operation of any battery storage system results when the data from every battery is used to tune how it responds to the grid commands, maximizing its responsiveness to the grid's needs and operating each battery in its optimal conditions.

Within the ESS, distributed controls will monitor and control individual subcomponents. For example, a cell monitor will communicate to a module monitor, which will communicate to a string manager, which may communicate to a bus manager, which communicates to the EMS. HVAC, PCS, and fire/safety control will also have their own independent controllers which tie into the EMS for cohesive site-wide management. Figure 6.7 shows a bird's-eye view of a grid-connected energy storage system.

6.3 Battery Chemistries

This chapter will explore a few of the most current or potentially important chemistries in the energy storage industry. According to the database of energy storage compiled by the U.S. DOE [3], the types of batteries used most extensively are lithium-ion, lead-acid, various flow batteries, variants of sodium, nickel, and zinc, and a few others (see Figures 6.8 and 6.9). According to this database, lead-acid was the first to be used in 1986, then nickel-cadmium in 2003, then lithium-ion in 2008, followed by many other technologies from 2000 onward.

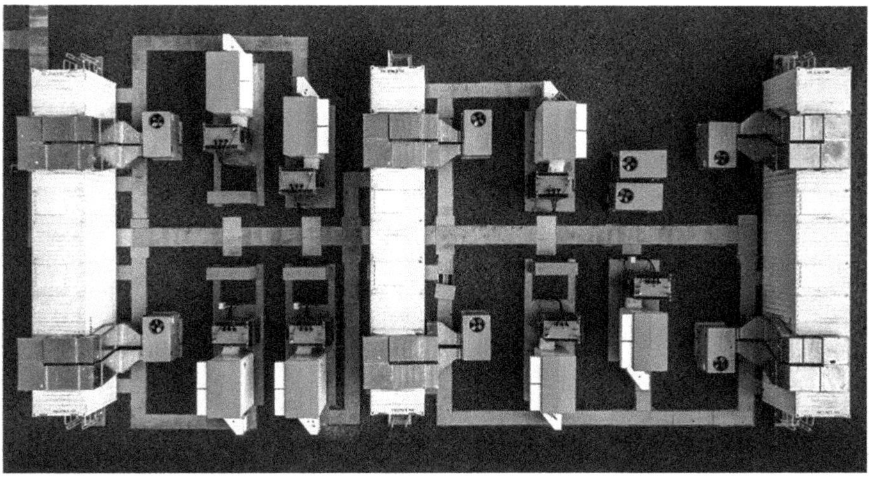

Figure 6.7 Example of a grid-connected battery energy storage system consisting of batteries, PCS, transformers, and cooling equipment (Source: NEC. Reprinted with permission).

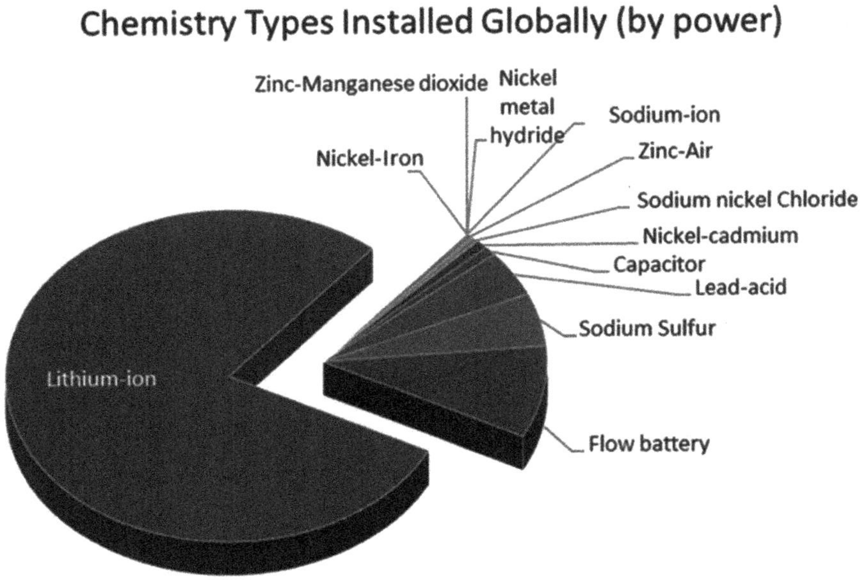

Figure 6.8 Chemistry types installed around the world by power, as of 2020 [3].

In the last decade the world has seen the most dynamic, fast-moving period of battery development since the early 1900s. Expect to see more improvements, more chemistries, and more updates to the text in this book, even as it goes to press.

The chemistries are presented mostly in the order in the which they came onto the market, but some have been reordered so the text flows better. Additionally, there may be others hidden in the shadows of trade secrets and prelaunch activities that were intentionally not included in this edition.

6.3.1 Lead-Acid

Lead-acid batteries have been around since the 1800s and the technology in general has seen many evolutions in its performance over the years. Its appealing qualities are its low cost and ease of use. This chemistry has been used to start almost every car that has existed since Henry Ford first did away with the hand-crank mechanism in his Model T, a

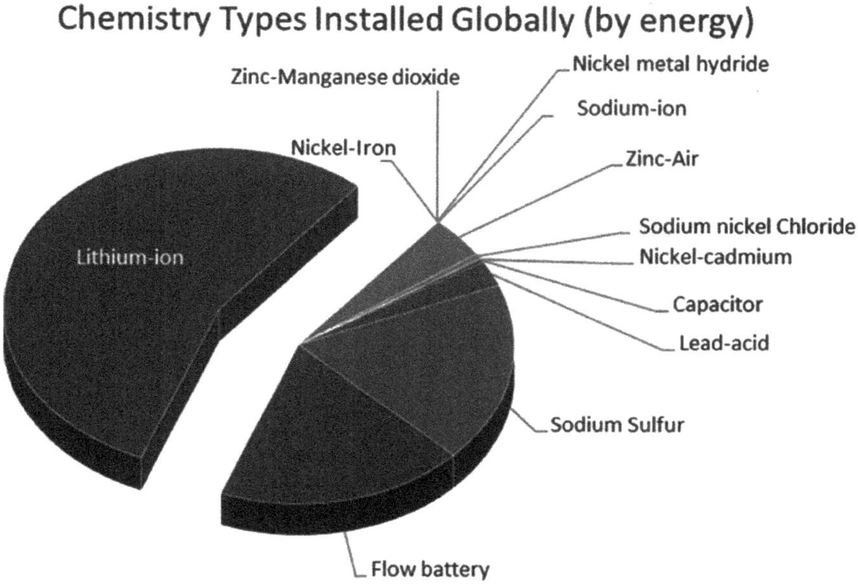

Figure 6.9 Chemistry types installed around the world by energy, as of 2020 [3].

source of many sprained wrists and bruised forearms. Lead-acid has also been used extensively in the telecom and uninterruptible power supply (UPS) industry, backing up the power to vital data and communications networks throughout the world. Up until the 2000s, almost every UPS ever manufactured used a lead-acid battery to provide back-up power when the grid power went out. In the telecom industry, rooms full of shelves containing lead-acid batteries provide days of back-up power to vital telephone and network infrastructure. Power is so critical to these networks, that fuses are sometimes omitted from the design because they introduce a minuscule, yet intolerable failure rate. Clearly lead-acid has served a worthwhile purpose in these industries.

Since the mid-1980s, 150 MW [3] of lead-acid energy storage has been brought online, of which 70 MW are still operational. Figure 6.10 shows the global lead-acid installations recorded in the GESDB [3]. Note the oldest ones have been decommissioned, and the vast majority of recent installations are under 1 MW.

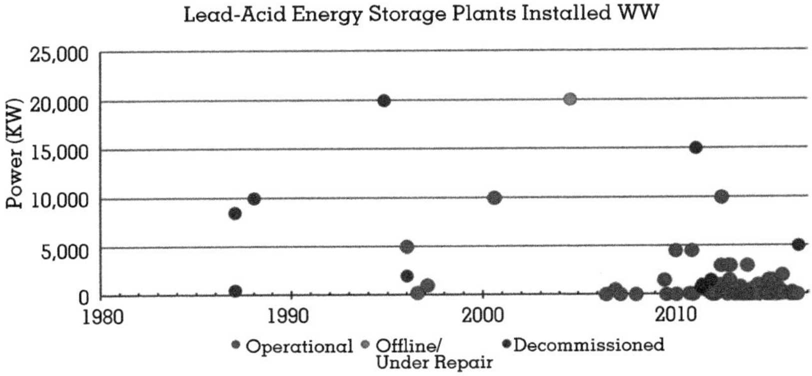

Figure 6.10 Lead-acid installations from 1986 through 2020 (listed by their power ratings) [3].

6.3.1.1 Chemistry

A lead-acid cell relies on a series of chemical reactions for its discharging and charging cycles. While charging, electrons are forced into the negative terminal by an external power source. The extra electrons release sulfate ion SO_4^{2-} into solution and cause lead sulfate $PbSO_4$ to be transformed into lead metal. At the positive electrode, electrons are drawn away from the terminal, causing the lead sulfate to break down into lead dioxide and free both H^+ and SO_4^{2-} ions into the electrolyte. During this charging process, the electrolyte becomes a more concentrated sulfuric acid. The state of charge can be determined simply by measuring the specific gravity of the electrolyte (a proxy for acidity). The string of electrochemical reactions that occur in this process are shown in (6.2), (6.3), and (6.4).

Negative plate charging reaction:

$$Pb + SO_4^{2-} \rightleftharpoons PbSO_4 + 2e^- \text{ (Discharged)} \quad (6.2)$$

Positive plate reaction:

$$PbO_2 + SO_4^{2-} + 4H^+ + 2e^- \rightleftharpoons PbSO_4 + 2H_2O \text{ (Discharged)} \quad (6.3)$$

The total charging reaction can be written as:

$$Pb + PbO_2 + 2H_2SO_4 \rightleftharpoons 2PbSO_4 + 2H_2O \text{ (Discharged)} \quad (6.4)$$

A physical representation of these reactions is shown in Figure 6.11.

Instead of ions shuttling between electrodes, the ions in the electrolyte react with the electrodes locally. Because ions do not have to travel the entire distance between electrodes for the reactions to occur, lead-acid batteries have extremely fast kinetics, leading to their high-power capabilities, albeit for short durations. In the discharge direction, sulfate ions in the electrolyte react with lead at the negative electrode and lead dioxide at the positive electrode to form lead sulfate. If the high-power pulse is extended for several minutes, however, the battery reveals its sluggishness. The rate of reaction kinetics with respect to the diffusion rate causes a concentration gradient of reactants orthogonal to the surfaces of the electrodes. This leads to a deprivation of ions at the surface of the electrodes and thus a lack of electrons needed to power the external load, manifesting itself as a voltage drop on the battery's terminals. That is one of the reasons why, when, after letting a starter battery sit for a few seconds after trying to start a stubborn car engine, one will discover new-found energy that seemed to be drained from the battery during the first try. While resting, the ions reequilibrate in the electrolyte and become available to react at the electrodes. When it is colder, the electrolyte is even more viscous, and they move even slower.

6.3.1.2 Performance

The formation of lead sulfate at both electrodes while discharging causes them to expand. When a sulfur atom and four oxygen atoms react with

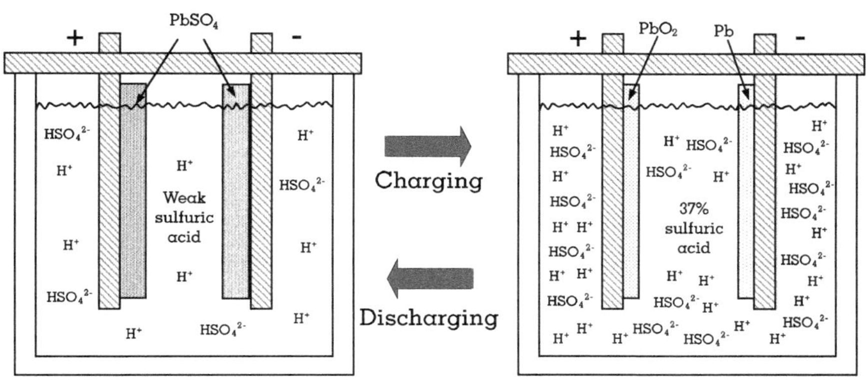

Figure 6.11 Simplified schematic of a lead-acid battery cell.

the lead metal, it can cause the electrode to expand as much as 50% when the cell is fully discharged. Knowing the construction of the electrodes can help to understand why this is a problem.

The electrodes are usually made of a solid lead-alloy plate cast in a rectangular grid pattern. A lead paste made with small lead particles and sulfuric acid is squeegeed into the grid structure. The small lead particles have lots of surface area at which reactions can occur. As the sulfate reacts with the lead at the surface of these particles, the particles grow and push against each other, encroaching into the space that the electrolyte previously occupied. Once full, the paste expands against the lead grid, sometimes with enough force to deform it. The paste even pushes itself out of the grid and against the separator between the plates. In flooded batteries, lead sulfate can fall off the plates and sink to the bottom of the cell. When the battery recharges, sulfate is freed from the lead particles still attached to the electrode and enters the solution again. The lead paste now occupies less space than when discharged, but it does not always fall back into the same shape as it started before the discharge. The plates may be bent, and the separator compressed, so the lead paste shrinks into a less than contiguous bunch of blobs, separated by an increased number of voids, gaps, and fissures. For the reactions to occur in either direction, electrons must travel between the terminals and through the materials in the electrodes. If those pathways are interrupted by an expanding network of gaps and voids in the lead paste, the reactions will be impeded.

This whole process manifests itself as a reduction in available capacity and increased internal cell resistance, albeit gradually, cycle-by-cycle. That is why lead-acid batteries are inherently weak at cycling deep discharges. Shallow discharges, like those used to start an engine or provide a brief pulse of power during power outages are not as damaging because the shallow growth of lead paste can be handled by its paste-like consistency. There are some variations of lead-acid batteries that are designed for deep discharges. They are touted as *deep-cycle* batteries and target the marine trolling and solar storage markets. The structure of the plates is more robust, the separator more elastic, and the electrodes contain more lead paste. All these design features improve the cyclability of the cell, in some cases up to 4,000 cycles [4], but come at the additional cost of more lead, more volume and more weight.

The reason lead makes up most of the materials in a lead-acid battery is that it is one of the few conductive materials that can withstand the corrosive environment of sulfuric acid. If copper or other metals are used in the cells, they are covered by a substantial layer of lead to protect them. The active materials of lead and lead-oxide are large and heavy. One lead atom corresponds to every two electrons transferred in and out of the cell and has an atomic mass of 207. In addition, extra lead is used in the cells to make them more robust and long-lasting, hence the weight and size lead to an energy density of around 35 Wh per kg and about 80 Wh per liter.

Lead-carbon batteries are a recent improvement that use nearly the same electrochemistry as a traditional lead-acid battery but adds some carbon to the anode, mixed in with the lead paste. Some designs even replace both electrode grids with porous hard carbon framework [5]. The carbon mixed with the lead will increase its conductivity and reduce lead sulfate crystallization on the anode surfaces during partial SOC conditions. This increases the power capability of the cell and enables it to stay at partial discharges longer without permanent damage. In addition, the increased amount of carbon that replaces an equal volume of lead, tends to decrease the weight of the battery, thus increasing its gravimetric energy density. Lead-carbon batteries have targeted partial SOC markets, such as frequency regulation, due to their higher power and mid-SOC tolerance. However, the user manuals are quick to point out that they cannot stay at partial SOC forever [6]. Periodic cycles to 100% SOC are required to refresh their conditions.

Lead-acid electrolyte consists of mostly sulfuric acid that reaches about 37% concentration when fully charged. In solution, the ions are H^+ and SO_4^{2-}. In a flooded-acid cell, this electrolyte fills the plastic jar and surrounds the positive and negative electrodes which are held in place by a mechanical structure. Most automotive starter batteries and many of the large central phone station battery jars are this type. When the battery is just about fully charged while it is being float-charged, it evolves a small amount of hydrogen and oxygen at its plates. The float voltage is just about the point at which the acid electrolyzes, and so a little bit of electrolysis happens at the top of charge. Older, flooded batteries have small vents in the top of their fill caps to let the hydrogen and oxygen out. That means that water is constantly being lost from those cells. That is why periodic maintenance is necessary to top off the electrolyte with distilled water in a vented lead-acid (VLA) battery. So-called "maintenance-free" lead-acid batteries partially seal the cells and force the gasses to recombine at their plates. That was the intent, but hydrogen is small enough that it can diffuse slowly through the plastic walls. Over the course of a few years, this leads to a gradual drying out of the cells and ultimate failure. In addition, if the battery is overcharged, (i.e., exposed) to a voltage much higher than its optimal float-voltage level, electrolysis will occur at such a fast rate, that the gasses will escape through the over-pressure valves in the caps, making a "bweep" sound as it escapes. These batteries are called valve-regulated lead-acid (VRLA) batteries because a partial seal acts as a pressure valve that only lets gasses out if they exceed a certain pressure, like 7 PSI, over the outside pressure. When the pressure is below that threshold, most of the gasses are contained and recombine with each other to form water again. These cells are sometimes referred to as recombinant cells for that reason. Some companies produce a catalyst cap that contains a catalytic material which causes the recombination to occur more efficiently, saving water and reducing hydrogen gas leakage.

In an absorbed glass mat (AGM) battery, a felt-like mat made of fiberglass holds the electrolyte in suspension between the plates. This mat allows the battery to be moved and reoriented without fear of the electrolyte spilling. In a gel cell, the electrolyte has additives that cause it to be more viscous and decrease its mobility. Many AGM cells are designed to be maintenance-free by including a relief valve, but because they do not have a reservoir of extra electrolyte as the flooded cells do, they tend not to last as long in service. However, an advantage of the AGM construction is that the mat and plates are lightly compressed together and tend to hold the lead paste in place as it expands and shrinks during cycling, leading to a higher cyclability over the flooded type. Figure 6.12 shows a representative design of the inside of an AGM battery.

6.3.1.3 Safety

Although touted as a safe alternative to lithium-ion, lead-acid batteries have had their share of excitement over the years including explosions, fires, and even thermal-runaway. Explosions can occur when the hydrogen that is generated during normal operation is not ventilated properly.

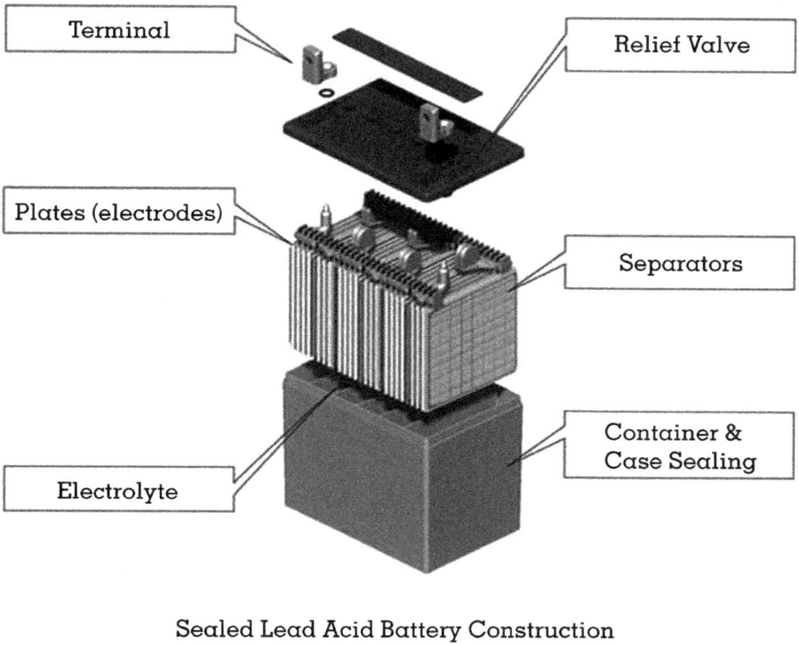

Figure 6.12 Lead-acid battery showing the various parts discussed in this section. (Source: Power-sonic. Reprinted with permission).

There have been cases where the roofs of buildings are blown off the building by the reaction of unventilated hydrogen gas with atmospheric oxygen. In one example, the ventilation system was accidentally shut off by a well-meaning occupant on their way out the door. Eventually, when the trapped hydrogen from a float-charging lead-acid battery accumulated enough, a spark from an unknown source initiated an explosion that blew a huge hole in the roof of the building [7].

That is why it is recommended that when connecting jumper cables to a vehicle's starter battery, that the positive terminal is connected first, and the negative terminal is connected to a point on the car's frame away from the battery. Too many times the sparks of the connecting jumper cable have ignited the hydrogen gas coming from the battery and caused a local explosion and operator injuries.

Although uncommon, a couple of the large lead-acid energy storage facilities in Figure 6.10 are decommissioned because of extensive fire damage. Questions remain relating to the cause of the fires, but it probably was not because of the cells themselves. Thermal runaway does occur, but its effects are not nearly as catastrophic as those of lithium-ion batteries. When an over-voltage is placed on a lead-acid battery, the water breaks down into hydrogen gas, oxygen, and releases heat. Heat raises the acid temperature and increases the likelihood of further electrolysis. This circular, positive feedback causes a high float current, high cell temperature, and excessive gassing [8]. If the ambient temperature is higher, this process will happen at a lower over-voltage level. During thermal runaway, the cell's temperature can get close to the boiling point of the acid and causes the plastic walls of the cell to become soft. The internal pressure of the newly formed gasses pushes on the walls of the battery and bloats them out, sometimes to the extremes shown in Figure 6.13. However, the maximum thermal-runaway temperature never exceeds that of the boiling point of the acid. When the gasses do vent during thermal runaway, they consist of hydrogen, oxygen, and a little hydrogen sulfide, so it will smell like rotten eggs. If the hydrogen is not

Figure 6.13 VRLA battery that was intentionally subjected to thermal-runaway conditions.

ventilated from the surrounding room, it can reach a concentration that can cause an explosion in the presence of heat or a spark.

6.3.1.4 Operating Lead-Acid Batteries

Lead-acid batteries have characteristics that require some unique operating protocols to maximize their service life. For one, lead-acid cells do best when they are fully charged after each cycle. In any lead-acid battery, electrolysis near top of charge slows the recharge process dramatically. As the SOC rises, the charge efficiency falls to near zero at top of charge. The current that enters the terminals, intended to increase the SOC, instead goes into creating H_2 and O_2. Theoretically, the time to recharge a lead-acid battery is infinitely long. Practically, it is five to ten hours after the voltage has reached its top of charge level. If a lead-acid battery is not float-charged at top of charge, it will not effectively clean off its plates. This accumulation of lead sulfate eventually crystallizes and locks up valuable capacity, reducing the battery's ability to store energy.

A charge cycle usually incorporates an initial stage where a fixed amount of power (CP) or current (CC) is injected into the cells to charge them to 80% to 90% of their full capacity. When the cell's terminal voltage reaches the maximum charge level, around 2.28V per cell, the process switches over to a constant-voltage (CV) mode in which the cell is held at a constant voltage until the current entering the cell falls to a plateau corresponding to its internal self-discharge rate. The latter stage is known as the float charge stage in which the cells are held at their optimal float voltage. The float charge could take hours to bring a cell up to its top of charge depending on the cell's design and temperature. Figure 6.14 is a representative time-plot of the charging process.

During a daily cycling protocol, it is important to bring the cells up to a full charge as soon as possible. If the daily regimen is void of this step, the cell will gradually lose capacity as a permanent layer of lead sulfate crystals grows on the surface of the electrodes.

Many large lead-acid batteries consisting of many cells in series are connected to power converters that monitor only the string voltage, not the individual cell voltages. Without monitoring the individual cell voltages, there is an increased risk that one or more of the cells in the string will be outside of its recommended operating range. Although lead-acid batteries have a natural tendency to self-balance when charged in a

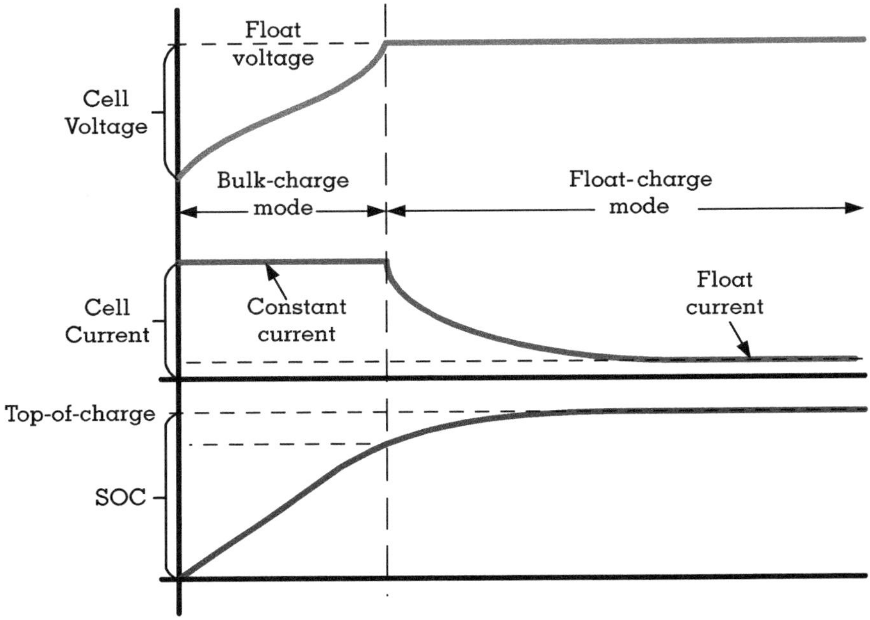

Figure 6.14 Representative voltage, current and SOC of a charging lead-acid battery vs time.

series string, such a mechanism can take a while to complete. Usually, UPS batteries do not have a problem with this issue because of the extended time between rare power outages. However, in a daily cycling application, cells can get out of balance with each other and not have a chance to self-equalize. The Chino battery plant experienced that very problem and discovered that they had cells operating with negative voltages during exhaustive discharges and others that ran with extremely high voltages during full charge cycles [4]. Eventually, they learned they needed to monitor each individual cell and regulate the power converters according to the state of each cell, not just the average of them.

Because of the damaging effects of lead sulfate expansion, lead-acid batteries' cycle life is highly dependent on their cycles' depth of charge. The deeper they are, the more damaging they can be to the cells' inner structures. One way to avoid deep discharges in regular operation is to significantly oversize the rated energy with respect to its daily discharged energy. Additionally, the EMS algorithms that respond to grid commands and dispatch them to the battery storage should be programmed to assign a high cost-factor to deep discharges so that it tends to avoid them, but not completely prevent them in case the grid needs a large amount of energy once in a while.

In addition to the main electrochemical reactions of (6.2), (6.3), and (6.4), other reactions occur on a continuous basis, intentional or not. When a cell is overcharged, not only does electrolysis happen, but reactions attack the lead structures on the positive plates, decreasing their lifetime. Conversely, when a cell is undercharged for long periods of time, not only does sulfation occur, but reactions attack the lead structures of the negative plate, decreasing its service life. In between, there is an optimal float voltage that maximizes the cell's service life, or said another way, decreases the rate at which deleterious reactions destroy itself from within. The rate of these reactions increases with temperature. In fact, it is estimated that the cycle life and service life are both reduced by half for every 10°C rise in its operating temperature. Therefore, it is vital that the cells are maintained between 20°C and 25°C. In addition, the optimal float voltage is a function of the cell's temperature. Higher temperatures require a reduction in the float voltage, therefore well-designed battery chargers will adjust their charging voltage according to the ambient temperature.

6.3.1.5 Examples of Lead-Acid Energy Storage in Service

As shown in Figure 6.10, there have been many lead-acid energy storage installations over the years. The first one was a 500-kW battery in California. The second, in Chino, California, was at the time the largest grid battery in the world at 40 MWh. The batteries occupied two buildings and a total of 4,500m^2 (shown in Figure 6.15).

There were eight series strings, each containing 1,032 2V cells that produced over 2,000V. The strings were combined and fed into one grid-tied, 10-MW power converter from ABB. The cells were the flooded type, so two features of this facility were necessary. One was a semi-automatic watering system that dripped water into the cells whenever the electrolyte levels were low. The second was a ventilation system necessary to keep the hydrogen from accumulating in the building. In addition, spark arrestors, specific gravity sensors, and electrolyte mixers were installed at each cell. The watering system was beset with problems from

Figure 6.15 The site of the largest lead-acid energy storage facility in Chino, CA [4].

the beginning, and the laborious chore of filling each jar manually had to be performed more often than originally expected. In the first year of operation, this large facility faced other issues such as roof leaks, fires, ventilation failures, imbalanced cells, PCS failures, and others causing the operating availability of this facility to range from 23% to 87%. The round-trip efficiency of the plant was around 72% including the PCS, transformer, batteries, and cooling system. One of the biggest contributors to the low efficiency was the recharge process. Typically, 109% of the battery capacity was put into it during each recharge cycle because of the typical lead-acid charge inefficiencies [4, 9].

The cells from Exide were guaranteed to last 2,000 full cycles, but independent testing indicated they could reach up to 4,000 cycles. This impressive ability was due to their robust design, including extra electrolyte in a flooded format and plates that sandwiched layers of absorbed glass mat and porous rubber. This flexible separator kept the active material in place during deep cycling. The plates were alloys of lead and tin (positive) and lead and calcium (negative), tuned to minimize corrosion over their lifetime. The Chino battery system was commissioned in 1988 and decommissioned nine years later in 1997. It served to demonstrate the possibility, although perhaps not the viability, of energy storage using batteries and pioneered the way for others to follow in the coming decades.

6.3.2 Nickel-Cadmium

Many people cringe at the thought of a large nickel-cadmium energy storage system. Their negative experience with it is probably with the low-cost consumer toys, tools, toothbrushes, and even short-lived laptop batteries back in the day when nickel-cadmium ruled the consumer world of batteries. Nickel-cadmium batteries were cheap but had a terrible habit of losing power too quickly during a cycle, and over their lifetime. On top of that, there was the clause in the instruction manual, which stuck in the minds of many people to this day, that they had to fully discharge their tool to zero every time they used it to prevent the dreaded "memory" effect. That is where the cell would remember its previous pattern of repetitive short depth of discharges and deliver only that shortened discharge during subsequent discharges.

It turns out that these limitations were caused by a combination of the chemistry and construction of the cell. The memory effect, for example, is only experienced with sintered plate, sealed Ni-Cd cells, used in many consumer products. However, this behavior is not consistent, and some cells never even experience that effect. In fact, large format, vented, pocket-plate Ni-Cd cells do not experience this effect at all and have been successfully performing in service for over 20 years as of this writing [10].

6.3.2.1 Chemistry

Ni-Cd cells consist of a nickel-oxyhydroxide NiO(OH) cathode, a cadmium anode, and an aqueous potassium-hydroxide (KOH) electrolyte. The charge carriers that move inside the cell between electrodes are negatively charged hydroxide ions (OH^-). While discharging, water and nickel-oxyhydroxide NiO(OH) at the cathode accept electrons from the external circuit, and convert to nickel-hydroxide $Ni(OH)_2$ and hydroxide ions, the latter of which enter the electrolyte. Meanwhile, an electron is removed from the anode and forces a hydroxide ion near the anode to react with the cadmium metal to form cadmium hydroxide $Cd(OH)_2$.

Negative Plate (Anode):

$$Cd + 2OH^- \rightleftharpoons 2Cd(OH)_2 + 2e^- \text{ (Discharged)} \tag{6.5}$$

Positive Plate (Cathode):

$$2NiO(OH) + 2H_2O + 2e^- \rightleftharpoons 2Ni(OH)_2 + 2OH^- \text{ (Discharged)} \tag{6.6}$$

Hydroxide ions react at both electrodes, (6.5) and (6.6), so the net effect is a movement of hydroxide ions from one electrode to the other in a half-cycle. Like the aqueous lead-acid battery, the main reactions are not the only ones. During the charging cycle gasses are formed, especially near the top of charge. In a sealed cell, the excess oxygen initially builds pressure within the cell, but eventually the O_2 recombines to form hydroxide ions, thus keeping the KOH in equilibrium. Under severe overcharging conditions, the excess oxygen can build enough pressure to rupture the safety vent at the top of the cell. Flooded Ni-Cd cells on the other hand, lose this oxygen continuously while float charging, requiring an occasional refill of distilled water. In fact, large installations of stationary Ni-Cd batteries are usually outfitted with automatic watering systems that periodically drip water into the cells.

6.3.2.2 Operating Ni-Cd Batteries

Ni-Cd batteries have an average voltage of around 1.2V during a full cycle, but unlike other chemistries, their charge voltage is not necessarily monotonic. From a fully discharged state, the cell will start at a voltage below 0.8V and climb to 1.4V at the top of charge. If the charger continues to inject a constant current, the voltage will reach a peak level at around 1.5V to 1.6V, plateau, and then fall to another plateau and maintain that lower plateau for as long as the cell is being charged. Figure 6.16 is a representative time plot of this charging process.

The temperature rises precipitously as the voltage reaches its first plateau and continues to rise as the cell's voltage falls to its second plateau. This temperature rise is a clue to what is happening in the cell at the top of charge. When the cell becomes fully charged, the positive electrode runs out of $Ni(OH)_2$ and cannot be further converted to $NiO(OH)$.

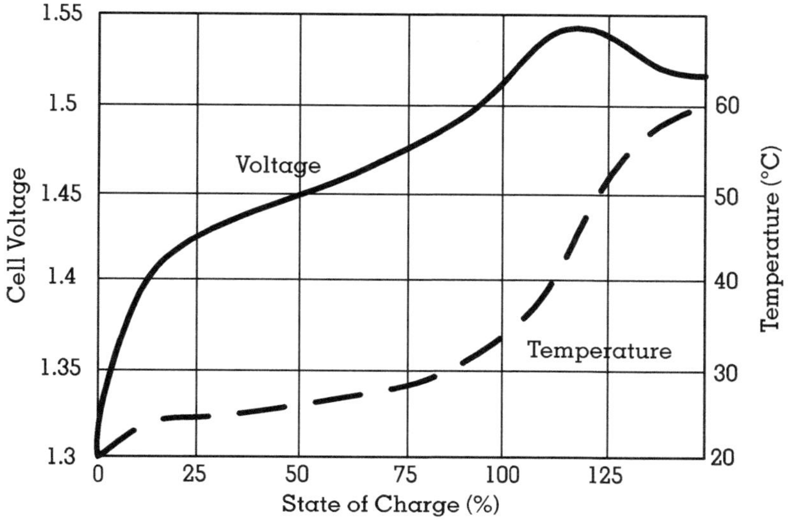

Figure 6.16 Representative charging voltage and temperature in a sealed Ni-Cd cell. (Notice the charging proceeds past the 100% SOC, because the horizontal access represents the accumulation of actual charge with respect to the capacity of the cell, not the relative SOC charge of the cell.)

So, the excessive OH^- ions lose electrons to the terminal and convert to water and oxygen gas. The oxygen disperses in the sealed cell, through the permeable separator and, upon receiving electrons from the negative terminal, reacts with water to form hydroxide ions. If too much oxygen is generated it will react with the newly charged cadmium metal and oxidize it, forming $Cd(OH)_2$, which is why a slight reduction of voltage occurs just after the cell reaches full capacity. Because this is an exothermic process, the cell gets warm at the end of a charge half-cycle. If, at this point, the charge current is reduced to a low level (usually around C/10), the cell's reaction rates can be maintained in equilibrium, where the excess oxygen does not blow the vent, and the heat of these exothermic reactions can be dissipated. On the other hand, if the electrical power is higher than that which can be effectively handled by the cell, the cells can self-destruct from excessively high temperatures and pressures. Although Ni-Cd cells can withstand higher temperatures than typical lead-acid batteries, they degrade rapidly above 45°C.

The mechanism of exothermic overcharge reactions is ideal for balancing the state of charge of cells in a series string. If one or more of the series cells has more charge in it than others in the same string, they will reach the top of charge sooner, and as they do, they will dissipate the excess charge current imposed on them to heat, while the other cells in the string catch up to their state. As soon as all the cells in the string have achieved their full capacity, their overcharge reactions will settle to a low level, and the string will be ready for the next discharge. However, if not enough time is allowed to let this process conclude, the string could easily become imbalanced. Ideal applications for Ni-Cd batteries are where there are plenty of opportunities for the strings to be at top of charge for long periods. An example of such an application is the standby UPS application where power outages are infrequent.

Because voltage is not a reliable method to determine state of charge, Ni-Cd battery chargers often employ other means to detect their charge state. Temperature threshold, rate of change of temperature, and even negative voltage rate detection are just a sampling of the methods to detect end-of-charge in Ni-Cd batteries. Some low-budget chargers even apply a very small constant current into the cells that is high enough to charge the cells, but not high enough to overheat them. Saft, a manufacturer of flooded Ni-Cd batteries recommends a float-voltage regulated charge, set to 1.40V to 1.45V per cell, with a current limit of 20% their rated capacity. Under these conditions, cells will charge to about 80% in 5 hours and then take another 10 to 15 hours to fully charge. A chart of various charge voltage profiles is shown in Figure 6.17.

Another important aspect of Ni-Cd batteries is their self-discharge. Because the active materials of the electrodes are surrounded by the hydroxide ions in solution, there are occasional random reactions that cause the electrodes to reduce their state of charge. This is accelerated at higher temperatures (i.e., above 25°C), but at 25°C, cells can lose 10% in 50 days. The rate of SOC loss tapers over time and is not linear as shown in Figure 6.18.

Because of this self-discharge, it is important to float, or trickle charge the cells to overcome this constant loss of charge. The level of trickle charge current is proportional to the absolute value of the slope of the curves in Figure 6.18. Notice that the amount of trickle current is exponentially proportional to the temperature of the cell. In standby applications, it will behoove the operator to maintain a cool environment around the cells to reduce trickle charge power consumption.

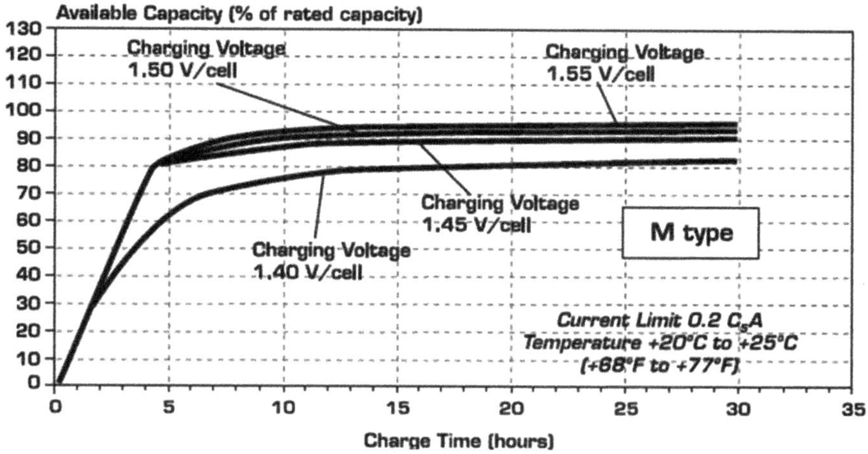

Figure 6.17 Charging voltage recommended by Saft for one of its Ni-Cd products. (From: Saft Batteries. [11].)

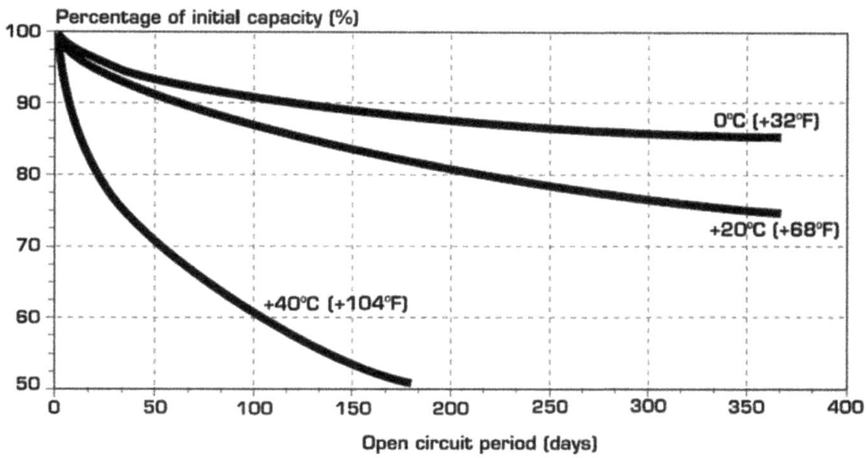

Figure 6.18 SOC of a Ni-Cd cell if left off the charger at various ambient temperatures. (From: Saft Batteries. [11].)

6.3.2.3 Performance

Large monobloc batteries are available up to over 1,000 Ah of capacity, that have a volumetric energy density of 50 Wh per liter, and a gravimetric energy density of 30 Wh per kg. This is slightly better than lead-acid batteries. Smaller, cylindrical cells have been developed with high surface-area materials, such as sponge metal and sintered plates, that have achieved up to 200 Wh per liter and over 60 Wh per kg. The high surface areas on the electrode materials also improve the rate capability of these cells, allowing them to discharge most of their capacity within 12 minutes.

One notable feature about a Ni-Cd cell, is that its terminal voltage while discharging is relatively flat for most of the discharge. It starts at 1.4V where it was float charging, settles to 1.2V, and is relatively constant until the end of its discharge, at which point the voltage rapidly falls below one volt. This flat voltage profile makes for a very difficult SOC meter. The high self-discharge and temperature dependent internal resistance further compound the difficulties in determining an accurate SOC.

The rate of Ni-Cd battery degradation increases at higher temperatures, but they do not degrade nearly as fast as lead-acid batteries. Whereas lead-acid batteries will experience twice the capacity degradation for every 8°C to 10°C above 25°C, a Ni-Cd battery will suffer an additional 20% of accelerated aging for every 9°C above 25°C. Most of the failures of a Ni-Cd cell occur because of a short-circuit across the separator. Cadmium tends to migrate through the separator and form small (i.e., high-impedance) bridges between the electrodes. This bridge causes excessive self-discharge that cannot be overcome by the external charger, and ultimately a dead cell.

The potassium hydroxide electrolyte (KOH) is toxic, corrosive, and a skin irritant, but it does not cause damage to a variety of metals and plastics used in the construction of the cells. For example, some cells use a steel structure to hold the active ingredients in the cell, which does not corrode while immersed in the KOH. The use of steel structural components enables a longer lasting service life and a more robust design. Typically, a Ni-Cd battery will last for over 1,000 cycles, and in a standby application, operate for 20 years or more, with proper maintenance.

6.3.2.4 Environmental Impact

The nickel and cadmium metals are fully recyclable, but the cadmium is toxic to humans and wildlife when it leaches into ground water and/or when burned. Responsible manufacturers like Saft have a comprehensive recycling program included in the purchase price of the cells. Nevertheless, because of its toxicity, cells containing cadmium are banned by the European Union for most consumer industries, but it is still allowed in certain medical and industrial applications.

6.3.2.5 Examples of Ni-Cd Energy Storage in Service

The first and largest Ni-Cd energy storage system was installed in 2003 by Saft and ABB in Alaska [12, 13]. Its purpose was primarily to perform spinning reserve. In this application, if a major generating source or tie line is suddenly interrupted, this BESS can immediately provide power to the grid to support the load for short durations (i.e., 5 to 15 minutes). The rated energy is 6.5 MWh at 26 MW output and 3.8 MWh at 46 MW output. The BESS was situated in a remote region in Alaska which suffered from large voltage variations due to the length of the 138-kV transmission line to the main grid. When large local loads turned on and off, the grid voltage fluctuated, causing power quality issues for other customers. The BESS was designed to respond to voltage, frequency, and grid disturbances, both planned and unplanned, and inject or absorb power to stabilize the local grid conditions. Since its installation in 2003, the BESS has responded to an average of 50 grid events per year [10], which includes major grid outages, tie-line disruptions, and unplanned loss of generation. During each event, the BESS successfully shielded its customers from an otherwise inevitable power outage. Although it only provides 15 minutes at 26 MW, this is long enough for the local utility to start up emergency power generators to sustain the loads through longer outages.

Remarkably, this BESS has a single PCS rated at 46 MVA made up of four modules per phase, driving three single-phase transformers. Four separate strings of 3,440 series cells deliver a dc bus voltage from 3,440V to 5,200V and can deliver over 12,000 amps to the PCS for short durations (see Figure 6.19).

The batteries themselves are rated to operate between 20 and 30 years, and the electronics are rated for 20 years. However, after 18 years of service, the system is showing its age and maintenance operations on the controls are increasing [14]. GVEA, the cooperative utility that owns and operates this BESS, is looking for options to upgrade in the next couple of years.

Figure 6.19 Inside the GVEA 27-MW Ni-Cd energy storage system in Alaska. (Photo credit: GVEA. Reprinted with permission).

Another energy storage installation using Ni-Cd batteries was installed in 2010 on the island of Bonaire [3]. The island's utility uses this 3-MW facility to help it achieve 100% use of renewable energy for its community's power needs. This utility has since added additional, larger lithium-ion batteries in the same grid.

6.3.3 Nickel-Metal Hydride

Nickel-metal hydride (NiMH) cells were first invented in 1959 by Dr. Wolf Vielstich at the Battelle-Geneva R&D Center. He proposed storing hydrogen interstitially between metal atoms. Unlike lead-acid and nickel cadmium cells, the electrode materials do not expand nearly as much when it has fully adsorbed the hydrogen atom. Originally, Dr. Vielstich used a simple nickel alloy with zinc or aluminum as a means to adsorb the hydrogen in storage. This and other early metal hydride materials were prone to fracturing and oxidation during the cycling process. In subsequent decades, refinements to the chemistry used different combinations of lanthanum, cerium, titanium, vanadium neodymium, praseodymium, nickel, cobalt, manganese, aluminum, zirconium, chromium, iron, and magnesium with varying degrees of success.

6.3.3.1 Chemistry

In the NiMH chemical reaction formulas, the alloy composed of two to five metals is symbolized as "M" and forms the negative electrode. The positive electrode is a nickel-oxyhydroxide and the electrolyte is potassium-hydroxide like that of the Ni-Cd cells.

On the negative electrode (anode):

$$OH^- + MH \rightleftharpoons H_2O + M + e \text{ (Discharged)} \tag{6.7}$$

On the positive electrode (cathode):

$$NiO(OH) + H_2O + e \rightleftharpoons Ni(OH)_2 + OH^- \text{ (Discharged)} \tag{6.8}$$

What is noteworthy about these reactions are that the water and hydroxide ions are present in both. When charging, a positive hydrogen ion jumps out of the cathode, and bonds with a nearby hydroxide ion to form water. Water moves toward the anode and reacts with the metal matrix to form MH leaving a hydroxide ion to balance with the one used up at the cathode. A single hydrogen atom becomes nestled among the large metal atoms. Simply put, water moves in one direction and hydroxide ions move in the other, resulting in the net effect of positive hydrogen ions moving from one electrode to the other and being either stored or released from the metal hydride in each half-cycle.

Like Ni-Cd cells, NiMH cells can endure overcharge conditions as long as the rate of current is low enough to enable internal recombination and heat dissipation. If the charge rate exceeds C/10 while the cell is fully charged (for most cell designs), heat and gasses will accumulate and damage the cell.

6.3.3.2 Performance

The performance of the NiMH cell is very much dependent on the mix of metals in the anode. The primary factors that affect performance are the ability of the metals to adsorb hydrogen at the required rates, the amount of hydrogen per unit volume, and the ability of the metal matrix to remain intact after cycling. The latter trait has been the most difficult and the reason it took many years to find the right combination of metals that could cycle as good as or better than nickel-cadmium. Stanley R. Ovshinsky and his team at Energy Conversion Devices (ECD) developed a superior combination of transition metals which they patented and was subsequently licensed by every NiMH battery manufactured.

At his weekly technology roundtable in February of 1982, Ovshinsky proudly demonstrated a working beaker-prototype of a cell using ECD's latest anode materials powering a fan. When Ovshinsky said that "Someday this technology will power electric vehicles," Subhash Dhar, asked him, "Who is going to do that?" Ovshinsky pointed across the table at Dhar and said, "You, my friend. You will do it." (Subhash Dhar, personal communication.) Subhash went on to become president of Ovonics Corp., a newly formed joint venture between ECD and American Natural Resources (ANR). He, with eight other scientists created the world's first commercially manufactured, sealed, sub-C sized, NiMH cell using ECD's latest mixture of transition metals. It was in many ways superior to its Ni-Cd cousin. Although it had similar voltage levels and patterns during charging and discharging, it was lighter and twice as energy dense, and it did not contain any toxic heavy metals. One of the downsides was that it had a higher self-discharge, but that was fixed by Sanyo in the 1990s with their Eneloop product. Other improvements in the materials of NiMH call design increased the energy density to as high as 300 Wh per liter by the mid-1990s. Many laptop manufacturers started using NiMH batteries because of these superior properties throughout the 1990s, Europe banned the use of Ni-Cd in favor of NiMH in portable applications, and in the mid-2000s half of all portable rechargeable batteries were NiMH.

The major problem with NiMH was lithium-ion. Soon after, Toyota started using NiMH cells in its Prius hybrid automobile and GM experimented with them in its trial EV1 program, lithium-ion batteries showed more promising energy density, lower self-discharge, and higher voltages. Consumers demanded more run-time for their devices, and lithium-ion batteries showed higher performance in that regard from the time they were invented in 1992 and made commercially available in the late 1990s. As good as NiMH was, lithium-ion was better in key market-differentiating respects.

6.3.3.3 Examples of NiMH Energy Storage in Service

In the DOE database of energy storage installations around the world [3], there is only one NiMH battery installation, a small 300-kW, 15-minute system used on a small island microgrid in Japan. It was installed in 2013 to help integrate the chaotic wind energy power profile and help stabilize the small island grid during variable wind conditions.

6.3.4 Lithium-Ion

In the 1970s, Stanley Whittingham introduced the concept of storing lithium ions between layers of other atoms in materials such as a disulfide. His first attempts at a lithium-based battery used a solid lithium anode, which proved to be a safety hazard, so productization was put on hold until a safer design could be realized. Important work by John Goodenough, Akira Yoshino, and J. O. Besenhard introduced ideas of a battery in which lithium only exists as an ion inside the cell. Sony introduced the first lithium-ion batteries in 1991, using the name lithium-ion to distinguish it from earlier attempts using lithium metal anodes. Their superior volumetric and gravimetric energy densities launched this technology into the forefront of portable electronics, transportation, and energy storage markets worldwide. The many different formulations, cathode and anode materials, and cell designs have allowed this chemistry to be used in applications from tiny earbuds to 500-MW stationary storage facilities. Cells could, and were, designed to either rapidly discharge in a few minutes or very slowly over the course of several days. Not only are lithium-ion cells smaller and lighter than previous chemistries, but they have much less self-discharge, do not emit gas under normal operating conditions, have a much more flexible SOC operating range, do not have a memory effect, and can be designed to have orders of magnitude higher cycle life. Despite these benefits, they have their limitations, as we will see later.

6.3.4.1 Chemistry

Unlike lead-acid, Ni-Cd and NiMH chemistries, lithium-ion cells are not powered by conventional redox reactions. Instead, lithium ions are intercalated between the layers of the atoms of the electrode materials (see Figure 6.20). Similar to NiMH, where hydrogen is stored among the

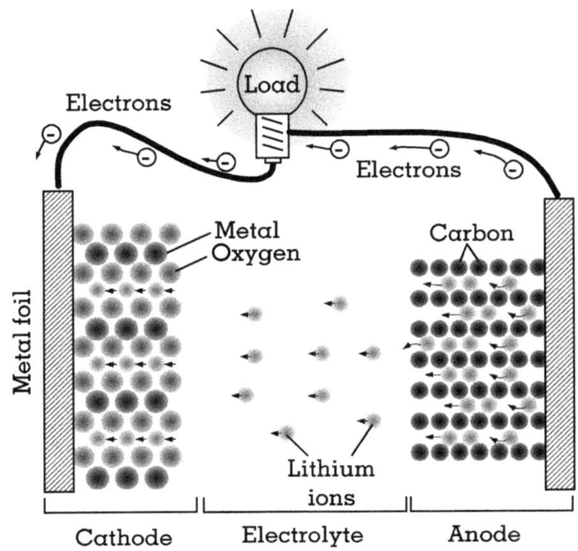

Figure 6.20 Lithium ions intercalate between the layers of either the cathode and anode while charging and discharging. Shown here are the ions moving in a discharging direction.

metal atoms of the anode, in the lithium-ion cell, a lithium ion is intercalated between the layers of the anode when charged and in the cathode when discharged.

The anode is most often made of a form of carbon, such as hard carbon, soft carbon, graphite, or a mix. There are many recipes of the carbon mixes, and they each have different effects on voltage, cycle-life, and efficiency of the cell in which they are used. In any case, the carbon particles are made of layers of a two-dimensional (2-D) carbon matrix, loosely bound together by Van-de-Vaal forces. When a cell is charging, its lithium-ions are inserted between the layers of 2-D carbon, attracted to this position by an influx of electrons from the external circuit. Other anode materials used in different variants of this chemistry include lithium titanate, tin cobalt alloy, and silicon, among others. Lithium titanate cells have been known for extremely high cycle life, up to 100,000 cycles, while silicon has been known to increase the energy storage density when added to the carbon in low ratios.

The cathode molecules are covalently bonded layers of metal (or phosphor) and oxygen, between which lithium intercalates as the cell discharges. Initially cobalt-oxide (LCO) was used exclusively, followed by manganese-oxide (LMO). Nickel-oxide (LNO) has the highest potential to store energy, but it is also the most thermodynamically reactive to abusive conditions. Mixes of those three have been extensively used and are known as NMC for nickel, manganese, and cobalt. First a 1:1:1 ratio of each, respectively, then 4:3:3, 5:3:2, 6:2:2, 7:2:1, and finally 8:1:1 have been used. Additionally, mixes of nickel, cobalt, and aluminum (NCA), and doped phosphate are also widely used. The different materials used each bring different attributes to the cells. For example, a high nickel content will yield a higher energy density, while cobalt will increase the power available from the cell and manganese will help reduce the cell's thermal volatility. As opposed to the metal oxide cathodes, phosphate-based cathodes are known for their benign thermal-runaway behavior, but cells with phosphate cathodes have up to 20% less energy density, and lower cell voltages than their oxide competitors. Additionally, phosphate by itself is not electrically conductive and requires either carbon coatings or a metal dopant, such as iron, to improve its rate capability. The most common form of phosphate-based lithium-ion batteries is an iron-phosphate known as LFP.

Lithium-ion battery electrolytes are usually a lithium salt compound such as $LiPF_6$ dissolved in an organic solvent such as ethylene carbonate. There are many variations of the electrolyte ingredients and additives that give it its characteristic behavior at various temperatures, voltages, and rates, depending on what application the cell was designed for. Each manufacturer closely guards their electrolyte recipe due to the years of investigation and testing required to develop winning formulations. The electrolyte is primarily responsible for allowing the rapid transport of lithium-ions from one electrode to the other during the charge and discharge half-cycles. It does not take part in the cycle-by-cycle reactions at either electrode, but it reacts initially with the anode particles during formation to get the cell ready for a long service life ahead.

Although initially the electrolyte was liquid, much research is being done on creating a solid electrolyte that can eliminate some of the undesirable aspects of flammable organic liquids. Some possibilities include polymers and ceramics. Although much progress has been made on reducing the ionic resistivity through these materials, the problem is that solid electrolytes don't have complete contact with all the surface areas of the electrodes like a liquid electrolyte does. This makes it difficult for ions to get in and out of storage and into the electrolyte. Whereas with a liquid electrolyte, the ions can readily diffuse through the liquid and access any surface in contact with the liquid.

6.3.4.2 Cell Construction

Lithium-ion cells are usually made from multiple layers of thin metal foils, coated with active materials (anode and cathode), and separated by an insulating separator. The many layers of thin foils increase the surface area and decrease the distance that the ions have to move, both of which increase the energy efficiency of the cell. Lithium-ion batteries typically have 85% to 98% round-trip energy efficiency, depending on their design, chemistry, and cycle power.

The active materials are finely ground powders of varying particle size. Small particles tend to enable higher power and longer cycle life, while larger particles tend to allow for a higher energy density. Some manufacturers mix the particle size to get the best of both power and energy. The powders are mixed with liquid and polymer binders, coated onto the foil, dried, and then compressed between rollers to the desired density, porosity, and/or thickness. Typically, the active material thickness is on the order of 100 μm. This process is tightly controlled in a high-quality production line because this step determines many important characteristics of the cell, regardless of the cell chemistry.

The separator is a multipurpose component of the cell primarily intended to keep the electrodes from touching each other. In addition, the separator is porous to allow liquid electrolyte to saturate it and act as a conduit for ions to pass between the electrodes. Some separators have multiple layers that perform safety functions as well. One layer may be a ceramic coating intended to deflect growing dendrites that may form under certain conditions. It also protects the cell from unintended foreign particles in the manufacturing process. In a lithium-ion cell, the smallest of conductive particles, invisible to the naked eye can breach the separator, and lead to a thermally active event. Some separators have a shut-down layer that melts at a critical threshold temperature [15], effectively shutting off the flow of ions between the electrodes. The idea is that if a short circuit were to occur across the cell, the cell would heat up, the shut-down layer would stop the flow of ions, and the cell would safely stop providing current into the short circuit. Other layers may be included to prevent the separator from shrinking in high-temperature conditions. There have even been attempts at providing a fire-suppressing layer in the separator. The separator must perform all those functions and be extremely thin, on the order of 20 to 25 μm [15], so as not to degrade the power and energy density of the cell.

In a cylindrical cell, an insulator, a separator, and both electrodes are rolled into a jellyroll, while in a prismatic cell, the layers can be either stacked or wound in an elongated jellyroll shape (Figure 6.21). Figure 6.22 show a pouch cell where the layers are stacked and vacuum-packed in a flexible pouch made of layers of polymers and aluminum.

The cell case is also an important component of the cell. Initially, lithium-ion cells were packaged in small cylindrical shapes, 18 mm in diameter and 65 mm long (hence the name "18650"). One disadvantage of this format is that they do not pack well in a rectangular enclosure,

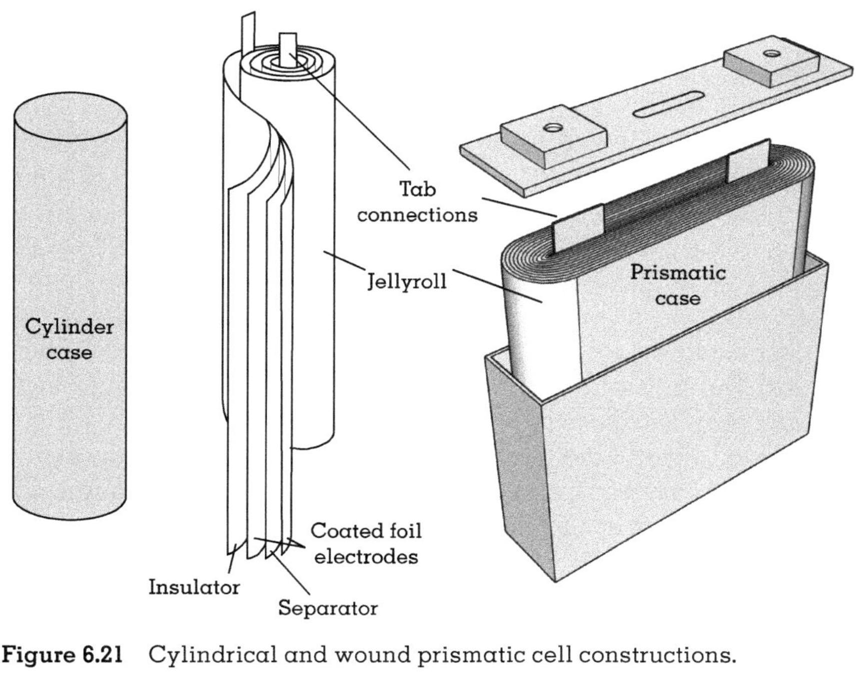

Figure 6.21 Cylindrical and wound prismatic cell constructions.

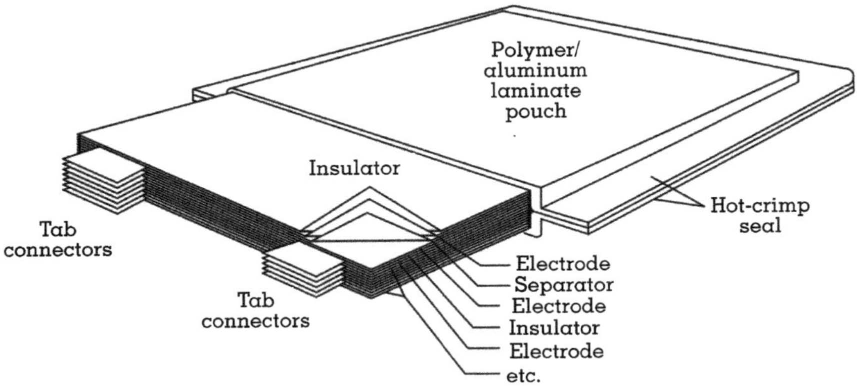

Figure 6.22 Stacked layers of a lithium-ion cell in a flexible pouch. Similar stacked structures are also used in some prismatic hard case cells.

and there will inevitably be airspace between the cells. If cooling is required around the cell's perimeters, this gap may be an advantage. Tesla Automotive chose to use a slightly larger 21700 cell for most of its vehicle batteries, and they are currently looking at a larger 4680 cell.

Prismatic cells tend to have more energy density in a box-like enclosure, because of less wasted space between cell walls. They also tend to be larger, thus enabling a higher energy density on a pack level. When cells expand during cycling and service life, the prismatic form-factor allows some expansion on the long sides, unlike the cylindrical, which tolerates little expansion. In any wound cell, if the expansion is not accounted for in the cell's internal construction, the layers could buckle, which leads to accelerated degradation at best and a short-circuit at worse. Prismatic cells tend to make for a simpler pack design because their enclosure is rigid and requires fewer mechanical parts. In both cylindrical and prismatic form factors, the cell case also includes overpressure vents and sometimes over-current protection.

Pouch cells are the most energy dense because their cell walls are the thinnest. That can prove to be an advantage and a disadvantage. Pouch cells are most commonly used in small mobile electronics because of their energy density and flexible formats. In larger packs, the thin walls require a mechanical structure to hold the cells in place and protect them from physical damage. However, the thin walls allow for better cooling if they are placed against a cooling plate. A good thermal design could allow the battery to function in high power applications and where the battery is constantly being cycled with little rest in between. Large, thick cells retain their heat so they require more time to cool between cycles and are more suited for one-cycle per-day applications.

6.3.4.3 Performance

It is not possible to generalize the characteristics of lithium-ion chemistry. The range of performance varies widely depending on materials selection, cell design, and even system design. The following section presents the general range in each of the performance categories and the design factors that affect them.

Energy

The practical energy density achieved by lithium cells ranges between 180 Wh and 900 Wh per liter (Figure 6.23), and 80 Wh and 300 Wh per kg. Lithium-titanate (LTO) has the lowest energy density because of

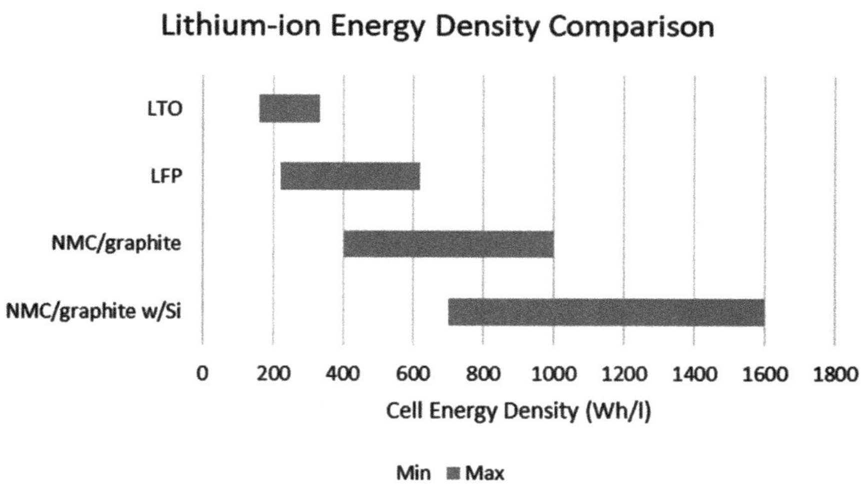

Figure 6.23 Ranges of lithium-ion volumetric energy densities for various lithium-ion sub-chemistries.

its low cell voltage of 2V to 2.8V. LFP is the next highest, with its voltage range between 2.8V and 3.6V. On the higher end, oxide cathode variants, such as LCO, NMC, and NCA have a voltage range between 3.0V and 4.2V and have correspondingly higher densities. Within that category, NMC with higher ratios of Ni (i.e., 8:1:1), will have even higher energy density because NiO_2 can store more lithium than cobalt and magnesium for a given volume. At the top of the range are cells with silicon mixed in with their graphite anode.

Silicon is the new horizon material for anode electrodes. It can store over ten times the lithium in its matrix, but it expands up to three times its volume when fully charged. The expansion causes the fragile Si structure to break apart and lose its storage capacity over a relatively short cycle-life (< 300 cycles). Lately, manufacturers have been adding a small amount (e.g., 5% to 10%) of silicon to the mix of carbon anode particles to increase cell capacity, but still retain a workable cycle-life (< 2,000 cycles). One area of research is embedding silicon in a flexible carbon scaffold that can absorb the expansion without breaking.

System Energy Density

One of the most common grid energy storage building blocks is a battery rack. There are many manufacturers of battery racks in the industry. The following are just a couple representative samples of such products. They are presented here to demonstrate the progress made in the industry in the last decade and are by no means a complete list. When A123 Systems started in the ESS industry, it produced a 24-inch wide rack that contained 35 kWh of energy storage, capable of delivering 140 kW of power. Because of its high power to energy ratio, it was targeted for frequency regulation markets. Later, they developed and sold 23 inch racks that contained over 100 kWh that could output 100 KW, and marketed them for higher energy applications, such as peak shifting. Both products were based on an LFP cell chemistry. Meanwhile, LG Chem, now LG Energy Solutions, created racks that could store 75, 120, 180, or 260 kWh depending on how deep they were. In addition, they evolved their NMC cell chemistry to target 4:1 energy to power applications to achieve over 300 kWh per rack. Samsung, SKI, CATL, BYD, Delta, and many others were also selling similar racks, all with slightly different power and energy capacities.

The latest LG Energy Solutions rack, shown in Figure 6.24 (on right) is 1m wide, 1.2m deep, and about 1.7m high that contains 330 kWh of energy (achievable at CP/4 discharge rate), and was used to populate the largest energy storage facility in the world at Moss Landing, CA. The rack's energy density is 150 Wh/l and 140 Wh/kg.

Power

The initial target market for lithium-ion cells was portable electronics, such as laptops and mobile phones. Because users desired many hours and even days of operation, the power rate of these cell designs was extremely low. Later, manufacturers designed cells that could discharge their energy in several minutes. Multiple aspects of the cell design contributed to this capability, including the electrode particle size, particle doping, electrolyte formulation, separator, conductor foils, number of tabs, and terminal configurations. Since some of these features also reduce the energy density, trade-offs were made when choosing which applications and markets they serve.

Figure 6.24 Examples of 35 kWh (NEC), 100 kWh (NEC), 75 to 260 kWh (LG ES), and 328 kWh racks (LG ES). (© 2022 LG Energy Solution, LTD. All rights reserved.)

Many of the features that promote higher power, will also add cost (i.e., thicker internal conductors), so manufacturers are wary of making a cell that is over-designed for its target market. Many manufacturers have multiple lines of products to accommodate different markets, so they can tune their costs appropriately. For instance, LG's JP3 cell is designed for 20- to 30-minute rates, while their JH4 cells are designed for 2- to 4-hour applications.

The system designs also impact its power to energy capability. Larger conductors, switching devices, and protective gear will allow the cell to discharge at higher rates. In addition, any heat generated in the cells must be removed from the system via heat sinks, air ducts, coolant pipes, pumps, and blowers. Some designs rely on the rarity of cycles to allow the cells to cool by themselves, while others actively move heat with a constant flow of coolant.

Cycle Life

Earlier lithium-ion products achieved 500 to 1,000 useful cycles and more cycles if discharged with shallow depth of discharges. Altairnano's lithium-titanate cell could perform over 10,000 cycles. A123 proved that smaller particle size not only improved rate capability but also cyclability. One of their earlier models cycled for 25,000 cycles before it rolled over at 50% of its beginning of life (BOL) capacity. As with power and energy, there is often a trade-off between cycle-life, energy density, and cost. Cells targeted for the automotive market generally have high energy density but only 1,000 to 2,000 cycles. An electric vehicle (EV) with a 500-mile range battery pack only needs 2,000 cycles to enable it to be a "million-mile" vehicle. Cells targeted for the energy storage market need to have at least 5,000 cycles to get past the 15-year, once-a-day service life. Fortunately, technology exists to tweak the ingredients and design to enable a product viable for either market. There are cells from mainstream manufacturers that can cycle 5,000, 8,000, or even 10,000 cycles for grid applications.

What is cycle life? Traditionally, battery cycle life was defined as the number of cycles that a battery endures before its capacity reaches 80% BOL. The 80% value comes from two sources. When a lead acid battery reaches 80% BOL, the remaining service life could take a drastic turn for the worse (i.e., its rate of capacity degradation accelerates unpredictably). This is sometimes referred to as the point of roll-over. Therefore, the useful service life of a battery ended when its capacity reaches 80% of its BOL capacity. The second source is the automotive market. If someone buys an EV with 100 miles of range on it, the consensus is that the car has outlived its useful potential when it can only drive 80 miles, or 80% of its BOL capacity. This is not a hard number, but given the precedent with 80%, it stuck with the automotive specification. In truth, lithium-ion batteries can continue to cycle without experiencing a roll-over capacity fade well after it has reached 80%. Some chemistries can get all the way down to 50% before rolling over, some 60%, and some 65%. Each manufacturer has their own point below which they cannot predict the performance decline. This is important in an energy storage application, in which batteries can continue to cycle until they cannot provide their required power. To compensate for loss of energy, old batteries can be augmented by adding new batteries throughout their service life, while the old batteries continue to cycle what energy still have within them. At some point, they will roll over and must be replaced entirely, but with careful site-planning this can be avoided.

Efficiency

Efficiency often goes hand-in-hand with the cell's rate capability, but there are other factors as well. A cell built with high power capability will have lower internal impedance, which reduces the i^2R losses in the cell, thereby increasing its efficiency. Lithium-ion cells have almost no self-discharge. If they leak faster than 3% in a month, that is considered high. There are no side reactions in a lithium-ion cell unless there is a problem with its condition or environment (i.e., an abusive condition). Therefore, every electron entering the cell moves an ion inside the cell. The losses in a cycle are mainly due to i^2R, (i.e., resistive). These resistances are both electronic, (i.e., electrons resisting flow in metal conductors), and ionic resistance, (i.e., ions resisting flow through electrolyte and electrode materials). Furthermore, for a given power and internal resistance, a cell with a lower terminal voltage will have more i^2R losses than for a cell with a higher terminal voltage. Therefore, LFP cells of similar power and energy rating as an oxide-based cell will have a point or two less round-trip efficiency.

That all said, lithium-ion cells have some of the highest efficiency of any battery chemistry, ranging from 85% to 98% round-trip. It is common to find metal-oxide cells having round-trip efficiencies of 97% to 98% when cycling at a CP/4 rate. LFP cells will, at the same rates, have between 94% and 96%. High power cells operating at 15- to 20-minute rates will see about 85% to 90% round-trip efficiency.

6.3.4.4 Safety

Safety is a hot topic regarding lithium-ion batteries. Many recent incidents involving consumer goods, electric vehicles, and stationary energy storage have highlighted the potential danger of this technology. Most of these incidents were caused or exacerbated by a thermal runaway condition in one or more of the cells in the involved battery pack. Thermal runaway can be caused by a number of factors, and once started is very difficult, and many times impossible, to stop. Thermal runaway is a process in which heat builds on itself and causes reactions which create more heat, and so on. Once an onset temperature is exceeded, thermal runaway can initiate, and the cell may catch fire or explode.

To get a cell to its onset temperature, the cell can be exposed to an external heat source, short circuited with an external short, internally shorted by crushing, puncturing, dendrites, or by a foreign particle, or it can be electrically abused by overcharging it.

There are several aspects of lithium-ion cell design that contribute to a highly destructive response to these thermal events. One is that when a cell is overcharged, the voltage can exceed the threshold over which the electrolyte will break down into gaseous compounds. Similar to how water electrolyzes above a certain voltage, the electrolyte will evolve into gasses when exposed to a high enough voltage. Temperature is also a contributing factor. Because gas evolution is an exothermic reaction, heat will be generated, which will cause additional gassing, which evolves more heat, and so on. Furthermore, the gasses are highly flammable, explosive, and dangerous. They comprise primarily hydrogen, carbon dioxide, carbon monoxide, and other volatile organic compounds like methane, ethane, ethylene, and others [16]. Under continued overcharge conditions, the gas pressure will build enough to blow through a safety vent built into the cell. The only safe aspect about this vent is that it prevents the cell from exploding uncontrollably from an even higher pressure later. When the gasses escape the cell, they mix with oxygen and, if the space around the cell is enclosed, an explosive gas mixture will be waiting for the next spark, heat source, or electrostatic discharge (ESD) event to set it off.

A cell can get overcharged for a number of unintended reasons. One of them can occur if individual cells are not monitored during the charge process. Consider a series string of cells being charged by a simple power supply connected to the most upper and lower points of the string (see Figure 6.25). If one of the cells has a higher SOC than the others, it will reach its maximum SOC before the others, and since all the cells have the same current (because they are wired in series), that cell will continue to be overcharged as all the others are brought up to their top of charge. This overcharging leads to an overvoltage, which leads to a breakdown in electrolyte, which leads to a catastrophic cell failure. That is why it is vital that BMS diligently monitor the status of each cell under its watch and be able to have influence over the current entering the string.

Second, the electrolyte in a lithium-ion cell is an organic solvent and, like paint thinner, is flammable. If the cell gets hot enough or is exposed to a nearby fire, the electrolyte will burn. When the cell is in the

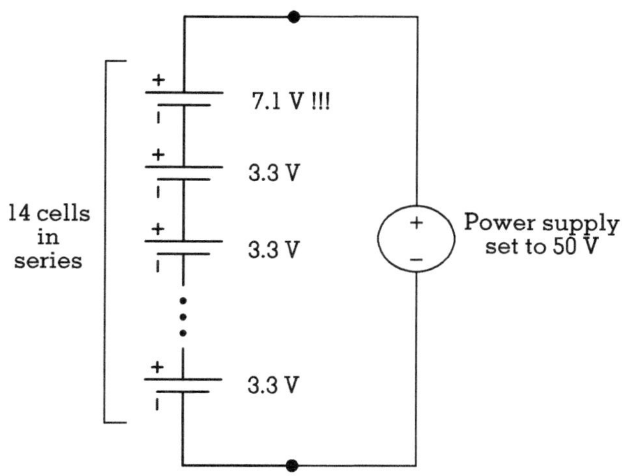

Figure 6.25 When a fixed voltage supply charges a string of imbalanced cells, some of the cells have the potential to become severely overcharged.

throes of thermal runaway, its temperature can exceed the electrolyte's flashpoint. This leads to the final dangerous aspect of lithium-ion cells, which is their metal-oxide cathode. Above a certain temperature, the metal oxide will enter its own thermal runaway. This one is much more rapid and hotter than that of the electrolyte and will release oxygen into the scene. The released oxygen fuels the flames that have already started by reacting with the other fuels released in the first stages of the thermal runaway event.

When a cell goes into thermal runaway, it starts out emitting lots of white smoke before it erupts into orange-colored flames, then it spews out white-hot gasses at nearly 1,000°C. Once this process has started, there is no way to stop it because the cell has its own fuel and oxygen. The best firefighting method is to stop the heat from spreading to adjacent cells by disbursing a cooling agent, such as water, on it and the surrounding cells. Figure 6.26 shows a chart of the temperatures of various cells induced into thermal runaway.

In these experiments, cells are heated up gradually until they reach a critical onset temperature, at which point the cell enters thermal runaway, and the temperatures spike to almost 900°C. Some cell chemistries, notably LFP, have higher onset temperatures and lower max spike temperatures. The NMC in these experiments is a 4.5:4.5:1 mixture, and spikes to almost 700°C.

A well-designed battery system with redundant overcharge protection, mechanical protection, and fully integrated over-current protection can prevent thermal runaway in most scenarios most of the time. However, in the case of an internal short-circuit, there is no amount of external protection that could prevent the likely thermal runaway. If a particle was introduced into the thin layers inside the cell during its manufacture, the BMS cannot stop it from shorting the cell. Therefore, a lithium-ion battery system is complete only when it includes a thermal suppression system capable of limiting the propagation from one cell to the rest of the pack.

Note: Some system integrators have at times chosen a let-it-burn philosophy wherein they intentionally leave out the fire suppression system and space each large enclosure arranged in a remote field to be far enough away from other batteries or flammable objects. However, there are so many other complicating variables that may not be accounted

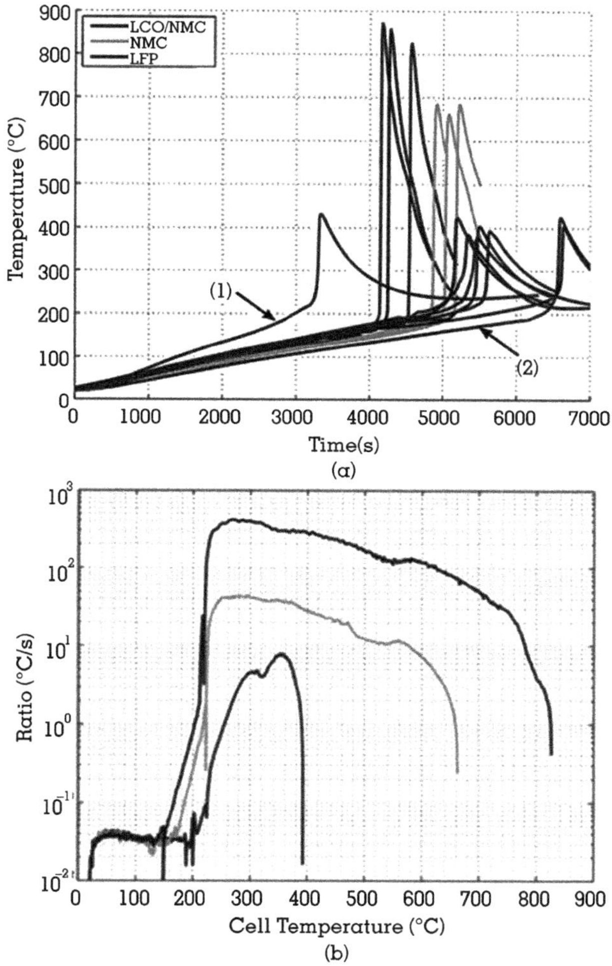

Figure 6.26 When lithium-ion cells are exposed to an external heat source or heat up from within, they can enter thermal-runaway at an onset temperature and experience rapid temperature spikes within seconds. (From: [16].)

for like prevailing wind and human behavior which are hard to predict during a 20-year service life.

6.3.4.5 Examples of Lithium-Ion Energy Storage in Service

According to the U.S. DOE database of energy storage installations, there are over 1,300 lithium-ion battery energy storage systems installed around the world [3]. This is probably a low estimate based on several known systems not included in the list. The list includes small 10-kW to 200-MW systems ranging from 15 minutes to 36 hours. According to several market reports, including iea.org [17] and Fortune Business Insights [18], there were around 5 GW of lithium-ion added to the global installed base, including behind-the-meter storage. The United States added 1.5 GW in 2020, which more than doubled its installed base in just one year. During 2020 and 2021, a few developers installed and contracted to install multiple projects in the 200- to 500-MW range per site ranging from 1 to 4 hours. The following paragraphs summarize a few of the milestone installations to show the progression of the state of the art in the last decade.

In 2007, AES Corp enlisted both Altairnano and A123 Systems to build the very first grid-scale lithium-ion battery energy storage systems. Altairnano completed their demonstration project in February of 2008 with two 1-MW/250-kWh systems based on lithium-titanate anode (Figure 6.27). It was demonstrated at the Indianapolis Power and Light facility and tested by KEMA. One of the 1-MW/250-kWh containers was moved to the PJM headquarters where it performed frequency regulation services until a hurricane damaged it beyond repair in 2012.

A123 Systems built a 2-MW, 560-kWh system in a single container and installed it at the AES Huntington Beach Power Plant in CA. It fulfilled its demonstration of following a frequency regulation profile and was moved a few times before settling in Puerto Rico, where it is now performing backup power services. Between 2008 and 2013, many other small demonstration lithium-ion sites were installed for demonstration purposes, sized between 5 KW and 5 MW, a few of which are worth mentioning.

In 2009, AES installed the world's first commercial lithium-ion grid energy storage system in Chile to help stabilize power for mining operations at the end of a 650-km transmission line (Figure 6.28). Like the Alaska BESS, this BESS supports the local customer loads and provides a key frequency response service for the broader system operator. The BESS, built by A123 Systems, is based on lithium-ion phosphate technology and can support 12 MW for 20 minutes from the top of charge. Typically, the battery is maintained near the midpoint of its SOC range so that it can support both positive and negative frequency deviations. More

Figure 6.27 Altairnano's first lithium system: 1 MW, 250 kWh used for frequency regulation services. (Source: Altairnano. Reprinted with permission)

Figure 6.28 12-MW, 20-minute lithium-ion system built by A123 Systems, installed by AES in the Atacama Desert, Chile. (Source: NEC. Reprinted with permission.)

than just a demonstration, this project increases the mining operations productivity and provides revenue-enhancing services for the region's power plant operators and is still in operation as of 2022.

Meanwhile, China was accelerating its electric vehicle and bus program, and to support their charging needs, created a 7-MW, 1.5-hour energy storage demonstration plant that could provide vehicle charging services without stressing the local power grid. The battery technology used was an iron phosphate from Wanxiang, the company that ultimately ended up buying A123 Systems out of bankruptcy in 2013.

Most of the projects up until 2011, of any significant size, were 1 hour or less in power duration, mostly due to the cost of the batteries and the higher value of the short-duration applications. However, in 2011, several multihour battery installations were completed by BYD (Build Your Dreams), Wanxiang, and ATL. They were 2, 3, 4, and 6 MW, with 4, 12, 16, and 36 MWh of energy, respectively. They were mostly designed to smooth the variable power from the many wind generators cropping up across the Chinese landscape. One notable project in China was a 1-MW, 36-hour system built by BYD for wind and solar renewable integration. Although 36 hours seems excessive, it demonstrated the flexibility in the power and energy scaling of a lithium-ion battery. To date, this was probably the longest duration battery energy storage project in the world. In the same year, AES installed another A123 Systems battery, which, at the time, was the most powerful battery energy storage system in the world at 32 MW. Like the other AES projects using A123 Systems batteries, it was only a 15-minute system and was designed to perform frequency regulation services in the PJM grid. Figure 6.29 shows the Laurel Mountain ES site's proximity to AES's 98 MW wind farm in the beautiful Appalachian Mountains of West Virginia.

In Maui, Hawaii, a 22-MW wind farm was being installed on a hill to provide the island with renewable energy and reduce their dependence on diesel fuel imports. Unfortunately, after the developer did their interconnection analysis, they realized that the wind farm's inconsistent power import would cause unacceptable power quality for the island residents. Wind surveys showed that the wind power had regularly fluctuated as much as 10 MW in a matter of seconds, and even 20 MW occasionally. Since the whole island sees about a 200-MW peak load, this variability would be hard to absorb by the few diesel generators scattered around the island. So, in 2012, MECO contracted Sempra to install an 11-MW, 20-minute battery, sized to react to 99% of wind conditions in a typical year (Figure 6.30). The battery was designed to smooth the rate at which the wind farm changed its power levels to the grid. This smoothing allows the diesel generators to ramp up/down their power output in response to changes in the wind power to provide a smooth and consistent power quality. The most encouraging aspect of

Figure 6.29 32-MW, 15-minute lithium-ion system built by A123 Systems, installed by AES in the hills of West Virginia. (Source: NEC. Reprinted with permission.)

Figure 6.30 11-MW, 4.3-MWh lithium-ion system built by A123 Systems, installed by Sempra to allow a 22-MW wind farm to provide renewable (and highly variable) power to a 200-MW island grid. (Source: NEC. Reprinted with permission.)

this project is that it enabled the utility to provide their customers with renewable energy and reduce their dependence on fuel in future years. Additional projects enabling both solar and wind energy followed in the footsteps of this project all over the Hawaiian Islands.

In 2014, Southern California Edison contracted AES to build a 100 MW/400 MWh BESS that does the job of a gas-fired peaker plant. A peaker plant is a relatively small power plant, compared to regular power plants, that is designed to respond to surges in power demands. They typically are not as efficient, but they can start up and respond to power demands quicker than the baseload power plants could. Battery energy storage systems can respond even faster than peaker plants and are much more efficient at doing so. In addition, they can even provide power absorption, whereas a fossil-powered peaker can only provide power delivery. In this bid, the AES battery energy storage solution proved to be more competitive than any other option, including gas-fired peaker plants. This was a turning point in the industry. Its size and function showed that batteries are not just considered a demonstration, but an economic, useful tool, enabling the electric grid to operate more effectively and efficiently. Contracted in 2014, Fluence built and delivered the Alamitos battery plant (Figure 6.31) to AES and commissioned it in 2019 [19].

In Texas, a large lead-acid facility of 36 MW was deemed end-of-life (EOL) after just four years of regulating variable wind power. Samsung bid and won the project in 2016 to replace the lead-acid batteries with a lithium-ion battery of 24 MWh. While most large lithium-ion bat-

Figure 6.31 100-MW, 4-hour lithium-ion battery, built by Fluence in a building for Southern California Edison. (Photo courtesy of AES. Reprinted with permission.)

teries were installed in an outdoor container-style enclosure, this battery was one of the few installed inside a large building.

In 2017 Tesla installed a 100-MW, 129-MWh energy storage system in Australia, which has since been scaled up to 150 MW, 190 MWh (Figure 6.32). It has reportedly saved its customers $75 million per year by allowing renewable energy to be used at any time the consumer needs it, not just when the sun is shining. An interactive graph on the Hornsdale website [20] vividly shows how the battery provides power to the grid when spot electricity prices are high and charges from the grid when energy prices are lower (Figure 6.33).

Notice there were times of negative energy pricing. At these times, solar energy was so plentiful that the utility was paying their customers to use it. At other times, energy supply was lower than demand, so energy prices were extremely high. This graph only shows a few days in recent history, but historical data from this website shows positive and negative pricing reaching as high in the tens of thousands of dollars per MWh and as low as −$1,000 per MWh. Price spikes like these indicate an extreme imbalance in a grid's power capacity and production and load requirements. Batteries, with their quick response times, are uniquely equipped to respond to these short price spikes and help stabilize the power grid.

Figure 6.32 100-MW, 129-MWh lithium-ion battery, built by Tesla in Hornsdale, Australia. (Source: [20]. Reprinted with permission.)

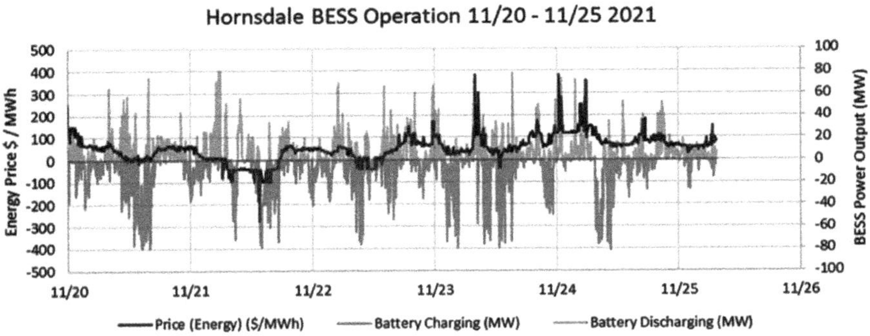

Figure 6.33 Tesla's lithium-ion battery, providing power to the grid as a function of spot electricity prices. (Data from: [20].)

According to the International Energy Association (IEA) [17], another 700 MW of lithium-ion battery energy storage was installed across the globe in 2014, followed by 1.2 GW in 2015, 1.8 GW in 2017, and 3.25 GW in 2018. Additions in 2019 were about the same as 2018, suffering a little setback from the coronavirus, but in 2020, IEA says that the world installed another 5 GW of battery energy storage, including behind-the-meter systems for home and business use. Other data sources, such as Wood Mackenzie [21], seem to indicate energy storage energy-to-power ratios are increasing as well, citing that in the United States, energy storage installations averaged around 3 hours for the first half of 2021.

For example, in August 2020, LS Power announced "the world's largest battery storage installation," which at the time was 250 MW/250 MWh, a 1-hour storage asset. It was built using LG Chem's NMC cell technology, integrated by NEC Energy Solutions, and installed by McCarthy Building Companies in San Diego, California. This BESS is an indoor design occupying five buildings, shown in Figure 6.34. However, from the onset, it was planned for future expansion. Figure 6.35 shows the inside of the building with double-stacked rack assemblies. As of this writing, they just finished installing another 500 MWh of energy storage, so the total run-time is now three hours at 250 MW of power.

Meanwhile, a few hours' drive away, also in California, the new world's record energy storage facility was built and commissioned between 2020 and 2021. The batteries of this 400 MW/1,600 MWh facility were built by LG Energy Solutions, and Burns & McDonnell provided

Figure 6.34 250-MW, 750-MWh lithium-ion battery built in a building integrated by NEC Energy Solutions. (Source: LS Power. Reprinted with permission.)

Figure 6.35 Inside the LS Power's Gateway 250-MW indoor BESS. (Source: LS Power. Reprinted with permission.)

engineering, procurement, and construction expertise. The system is colocated on the site of Vistra's existing Moss Landing Power Plant in Monterey County. The system was built in two phases. Phase I is located within a refurbished 1950s former turbine building. The thick concrete walls and floors were repurposed to provide mechanical stability and a large thermal sink to maintain the operating temperature of the batteries at a consistent level. Phase II is located inside a new building, constructed to house the battery system (see Figure 6.36). Phase III is being planned to increase it even further.

The 4-hour storage facility is the largest as of 2021, but is just one of many, present and future, in California to allow for 100% renewable energy to power the state's loads by 2045. Sites with multiple gigawatt-hours of energy storage will be common in the years ahead.

All around the world, cities, states, and countries are procuring gigawatt-hours of battery energy storage. In 2020, battery energy storage was just 5% of the total energy storage installed globally, where the remaining 95% was mostly pumped hydro. Although none of the industry analysts can agree on just how much energy storage is installed today, they all predict that the world will have five to ten times as much energy storage in another decade, with lithium-ion batteries outstripping pumped hydro in the coming years [17, 21, 22]. The U.S.-based National Renewable Energy Lab (NREL) published the graph in Figure 6.37, projecting energy storage installations up to 2030 in the United States. It

Figure 6.36 Inside the Vistra's Moss Landing 300-MW indoor BESS (Source: Vistra Corp. Reprinted with permission.) See also Figure 1.17.

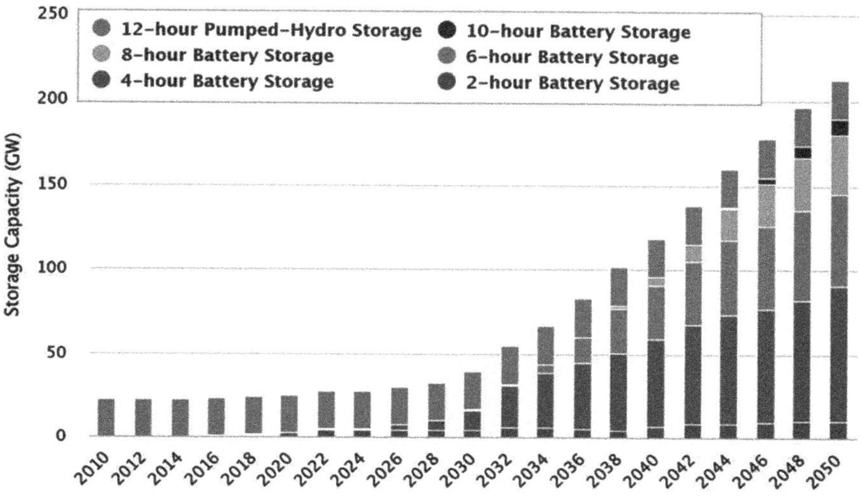

Figure 6.37 NREL's projected growth of installed energy storage in the United States. (From: [23].)

can be seen as a representative chart for how the world will expand its energy storage.

6.3.5 Sodium-Sulfur Batteries

This technology was originally developed by the Ford Motor company in the 1960s, but was later adopted by NGK Insulators, a Japanese company. Its high temperature maintains the sulfur and sodium in a liquid state, which vastly improves ionic conductivity, eliminates dendritic growth, enables electrode expansion, and increases cycle life over room temperature versions of this chemistry. NGK, a specialist in the beta-alumina ceramics used for the solid electrolyte, is the only company that has been selling this type of battery worldwide.

6.3.5.1 Chemistry

The chemistry of a sodium-sulfur cell is quite simple on paper. Molten sodium is contained in a ceramic cylinder surrounded by molten sulfur. The operating temperature of the cell is between 300°C and 360°C to keep these ingredients in a liquid state. When a sodium-sulfur cell discharges, positive sodium ions pass through the ceramic container walls and bond with sulfur atoms, while electrons flow from the sodium bath through the external circuit and into the sulfur cathode. During this process, the liquid level in the ceramic cylinder drops while the sulfur level increases. When the cell recharges, electrons are forced back into the sodium anode, which pulls the sodium ions back through the ceramic walls into storage again. The chemical reaction is as follows:

$$2Na + xS \rightleftharpoons Na_2S_x \text{ (Discharged, where } x = 3.3 \text{ to } 5) \tag{6.9}$$

6.3.5.2 Performance

Theoretically, the energy density of the sodium-sulfur chemistry is very high at 768 Wh/kg [1], but there are many practical impediments to achieving this. For one, since the liquid levels go up and down in the cell (see Figure 6.38), there are air spaces in the cells that are unavoidable. Additionally, the high-temperature components must be thick and robust enough to maintain a practical service life. At the pack level, thick insu-

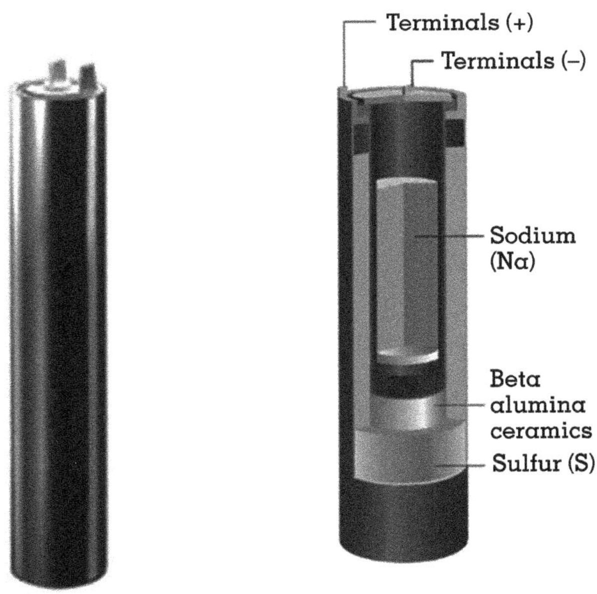

Figure 6.38 Molten sodium-sulfur cell. (Source: NGK Insulators. Reprinted with permission.)

lation needs to keep the cells at a high temperature without endangering service personnel. Additionally, to prevent thermal runaway propagation, cells are separated by a thermally isolating medium such as sand or other ceramic powder. These mechanical features contribute to a lower volumetric and gravimetric energy density. By themselves, sodium-sulfur cells have an energy density of 435 Wh/liter and 260 Wh per kg, a number competitive with LFP lithium-ion cells. After adding insulation and thermal isolation, the energy density of sodium-sulfur modules is reduced to about 100 Wh per liter. NGK has installed their 20-ft containers containing 1.45 MWh, double stacked to help improve their footprint in land-constrained projects as shown in Figure 6.39.

The ceramic cylinder containing molten sodium in a sodium-sulfur cell contains microscopic voids within the structure that selectively allow sodium ions to pass. The solid ceramic electrolyte is also very good

Figure 6.39 NGK battery containers double stacked. (Source: NGK Insulators. Reprinted with permission.)

at keeping the self-discharge to zero. However, the restrictive channels manifest as internal cell resistance. Hence the reason for the battery system having 75% to 80% round-trip efficiency. These losses are helpful in maintaining the cell's high temperature. Pack designers optimize the insulation around the pack to retain the heat of each round-trip cycle so that additional auxiliary power is not required to maintain the cells at their optimum temperature range. If a sodium-sulfur battery is not cycled daily, auxiliary power will be required to heat the inside of the pack to at least 300°C.

Because of a sodium-sulfur cell's relatively high dc impedance, they are most applicable for six hours of power duration (i.e., their power is rated for one-sixth of their energy rating).

Based on their actual field experience, NGK is confident that their products can last 7,300 cycles and 20 years of service life, whichever comes first. At the end of 7,300 cycles, they project their cells to have 64% remaining capacity in them.

6.3.5.3 Safety

Any storage system that concentrates power and energy in a small space can present a danger if that energy is inadvertently and uncontrollably released. Batteries are no exception, and, with regards to the sodium-sulfur chemistry, the failure mode can generate quite a lot of heat. In particular, if liquid sodium comes in contact with liquid sulfur, an instantaneous thermal runaway event will occur that will continue to rage until all the reactants are consumed. As with lithium-ion batteries, the preferred method to control the outcome is to prevent propagation from one cell to another. Otherwise, the thermal runaway can propagate to neighboring cells and, eventually, the whole pack can be consumed by a raging fire that could last for days. NGK's current design surrounds each cell by a thermal and electric insulator, a fire-resistance sheet, and a blanket of sand which absorbs the excessive heat and prevents neighboring cells from reaching their critical onset temperature.

6.3.5.4 Examples of Sodium-Sulfur Energy Storage in Service

According to the DOE's database of energy storage [3], there were 32 systems installed between 2005 and 2017, totaling 1.2 GWh of energy, but NGK claims there are over 200 installed worldwide, totaling 4 GWh.

Most of the installed systems are still in operation, quietly performing peak-shaving and bulk energy arbitrage daily. In any case, the largest is in Buzen City, Fukuoka, Kyusyu, Japan. Commissioned in 2016, it has 50 MW of power and 300 MWh of energy (see Figure 6.40). Batteries are housed in 252, 20-ft, 200-kW containers. The site was designed to smooth the effects of variable frequency and power resulting from nearby renewable energy sources, both solar and wind.

Another installation is worthy of mention for a couple of reasons. For one, it is a total of 108 MW and 648 MWh, it is distributed over 10 locations in the United Arab Emirates and, can be controlled to act in concert as one unit. It was designed to help defer the costs of utility infrastructure upgrades, perform energy shifting, and provide frequency regulation and other services for the entire UAE grid. Second, since it is a high-temperature battery, it does well in hot environments such as the UAE, unlike other battery chemistries, which must be cooled to 20°C to

Figure 6.40 50-MW, 300-MWh NGK battery energy storage in Buzen City, Fukuoka, Kyusyu, Japan. (Source: NGK Insulators. From: http://www.mitsubishi-electric.co.jp/news/2016/0303-b_zoom_01.html. Reprinted with permission.)

25°C continuously. The ten locations (one of them shown in Figure 6.41) were installed over eight years, from 2010 to 2018.

6.3.6 Flow Batteries

Flow batteries are a general class of batteries in which the positive and negative electrodes comprise liquids that are pumped from one or two storage tanks past an ion-exchange membrane, where they exchange ions or react in some way, then return to the tanks in an altered energy state. Electricity, which flows from one collector plate in contact with the liquid(s) through a circuit to the other side, either causes the liquids to gain or lose energy content as they are pumped along the surface of the membrane. Of the many variants of flow batteries, the main types are vanadium redox, zinc-bromine, and iron-iron.

6.3.7 Vanadium Redox Flow Battery

The all vanadium was invented by Maria Skyllas-Kazacos and her teammate at the University of South Wales in the mid-1980s. It has been developed into products by several companies including Uni-Energy

Figure 6.41 One of ten NGK battery systems in Abu Dhabi, UAE. (Source: NGK Insulators. From: https://www.energy-storage.news/uae-integrates-648mwh-of-sodium-sulfur-batteries-in-one-swoop/. Reprinted with permission.)

Technologies, StorEn, Sumitomo, Prudent Energy Systems, Avalon, H2, Gildemesiter, VRB Energy, Invinity Energy Systems, Largo, Rongke, Ashlawn, and even the Pacific Northwest National Laboratories. According to the GESDB [3], there are over 130 MWh of operational vanadium redox flow batteries (VRFB) as of 2020, and many more are already being contracted and planned in the next coming years.

6.3.7.1 Chemistry

Multiple ionic states of vanadium are used in the charge and discharge of a VRFB. Referring to Figure 6.42, there are two tanks, one containing V^{3+} and V^{2+} ions in an acid solution, and the other containing V^{4+} and V^{5+} ions (the V^{4+} and V^{5+} ions exist in the form of VO^{2+} and VO_2^+, respectively), also in an acid solution. When fully charged, one tank contains mostly V^{2+} ions, and the other contains mostly V^{5+} ions. The imbalance between these ions drives a voltage between the terminals when these

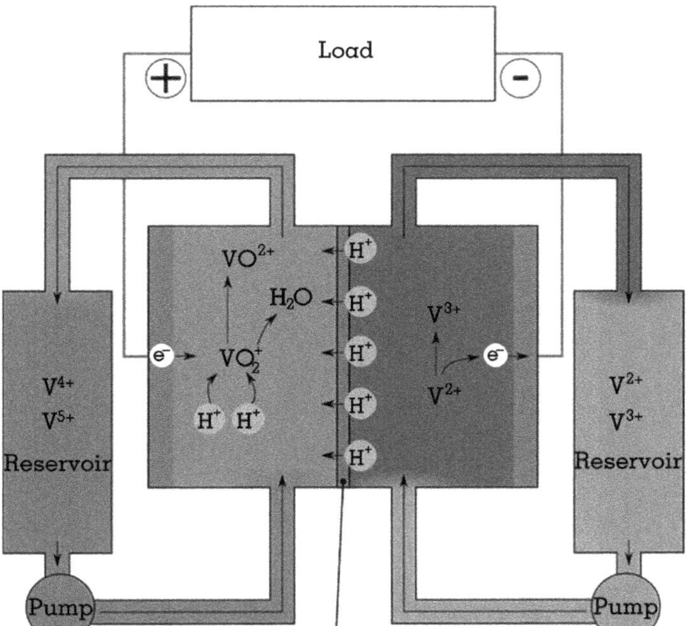

Figure 6.42 Pictorial representation of the vanadium redox reactions.

ions are pumped on either side of a membrane. If current is allowed to flow in the external circuit through a load, protons will flow across the membrane to force V^{2+} ions on the negative side to become V^{3+} ions and V^{5+} ions on the positive side to become V^{4+} ions. A discharged battery will have mostly V^{3+} ions in the negative tank (anolyte) and mostly V^{4+} in the positive tank (catholyte). The chemical reactions are summarized by the following formulas.

Negative electrode (anode) reactions:

$$V^{2+} \rightleftharpoons V^{3+} + e^{-} \text{ (Discharged)} \tag{6.10}$$

Positive electrode (cathode) reactions:

$$VO_2^{+} + 2H^{+} + e^{-} \rightleftharpoons VO^{2+} + H_2O \text{ (Discharged)} \tag{6.11}$$

6.3.7.2 Battery Design

There are two primary components of a flow battery: the stacks and the reservoirs. The stacks are built of layers of electrodes, fluid passages, and membranes. Each layer of the stack is an individual cell and has a charged to discharged voltage range of 1.7V to 0.8V, respectively. The greater the surface area of the membrane, the higher its current capacity. The reservoirs contain the liquid active materials, referred to as anolyte and catholyte. The balance of plant equipment consists of pumps, pipes, heat exchangers, power conversions, and controls.

One unique aspect of a pure, nonplating, flow battery is that the energy storage capacity can scale separately from the power capacity. Increasing the tank volume will increase the energy capacity and increasing the number of stacks or their physical size will increase the power capacity. Likewise, adding more series cells to the stack will increase the voltage and since the same electrolyte flows through each cell the stack's SOC is always balanced.

Because the electrolyte is somewhat conductive, there are limits to how many cells can be supplied from the same electrolyte reservoirs. Typically, manufacturers keep the voltage levels to under 100V per stack to keep the self-discharge to a tolerable level.

The membrane is similar to that used in proton exchange membrane (PEM) fuel cells; it allows only protons to pass through while blocking other ions and electrons from flowing. This design is not perfect, however. When charged liquids are on either side of the membrane, there is some crossover of unwanted ions, and as a result there is some self-discharge. Design trade-offs are made selecting membrane thickness; thicker membranes slowly self-discharge but also reduce power and efficiency and increase costs.

There are two approaches that manufacturers have taken to build VRFBs. One is to factory-assemble a complete enclosure that includes the stacks, tanks, balance of plant, and prefilled liquid tanks (see Figure 6.43). The other is to build the VRFB on-site with large, centralized tanks, dedicated power stacks, and conversion enclosures. The former installs much quicker and is less prone to site variabilities in cost, quality of workmanship, and schedule. The latter has the potential to be more energy-dense because of economies of scale.

Figure 6.43 Invinity's modular, stackable all-in-one VRFB system. (Source: Invinity Energy Systems. Reprinted with permission.)

6.3.7.3 Performance

Several manufacturers offer a modular format similar to that shown in Figure 6.43, in which the tanks and stacks are built into a container to simplify the installation process. Invinity's 224 kWh system fits into a 20-ft container that includes both electrolyte tanks and the stacks. Sumitomo also has a modular system in which some containers have the DC stacks, while other have the electrolyte tanks. Their 20-ft stack containers can provide 250 kW of power, while a 20-ft electrolyte containers provide 375 kWh of energy. They also have a version in which the electrolyte containers are replaced by large tanks and plumbed into the stack containers. VRB Energy is another VRFB manufacturer but has indoor products that can fit 6 MWh in a 310m^2 footprint.

Third-party testing [24] of VRFB products reveal a round-trip efficiency for full depth of charge cycling at around 62% to 75% depending on the rate and manufacturer, while partial depth of discharge cycling for frequency regulation is 75% to 80%. A large percentage of the losses occur in the pumps that transport the electrolyte from the tanks and through the stacks of cells. Typically, the pumps will draw about 5% to 10% of the full power rating of the stack. The effect of this number on a daily cycling profile can be reduced with some preplanning. Running the pumps while the stacks are not delivering or storing energy is not necessary for cyclic applications. For example, if the product is used for prescheduled load-shifting, (i.e., discharging) its energy from 5 PM to 7 PM every evening and recharging between 8 AM and 5 PM, the pumps can be turned off between the cycling periods. This will not only save auxiliary energy but reduce mechanical wear and tear as well. However, if the product is intended to respond quickly to power requests, then it may be necessary to keep the pumps constantly running so that the electrolyte is always in the stack and fresh. Otherwise, it could take a few minutes to start the pumps and bring fresh electrolyte from the tanks to the stacks to begin outputting full power.

The voltage range of each cell during a discharge will be between 1.4V and 0.8V, and up to 1.7V on recharge, so it will require a power conversion system that can accept such a wide range.

DC power capability is dependent on the area of the stack, the thickness of the membrane, and the concentration of freshly charged ions in the electrolytes. As the concentration difference near the top or bottom of charge decreases, the power capability in either direction will be reduced. Trustworthy manufacturers will provide the minimum power available over the whole SOC range but offer an additional peak power capacity for certain ranges of SOC. This may make them look worse than their competitors when comparing head-to-head, but it will not disappoint their customers when it comes time to run a full commissioning test.

One final note about VRFB, is that the electrolyte never degrades in its ability store energy. That means with a minimal amount of effort, the electrolyte from an older system can be transferred to a newer system. This is a good thing since the electrolyte is about 35% of the upfront cost of the VRFB, mainly due to the vanadium content.

6.3.7.4 Safety

The aqueous electrolyte is not flammable, and thermal run-away is not possible, so manufacturers point to their safety in comparison to lithium-ion batteries. However, an electrolyte spill would have to be carefully dealt with, and proper venting is required to mitigate hydrogen accumulation as sometimes happens with this technology. These precautions are typical for a lead-acid battery installation, and just shows that any concentration of energy in one form or another whether chemical, mechanical, or electrical must be respected.

6.3.7.5 Examples of VRFB Energy Storage in Service

As of 2021, there have been several dozen, small to medium VRFB installations around the world, sized between 5 kW to 15 MW, most of them under a MW. However, starting in 2020, China has been looking to build two large sites, with a combined total of 300 MW and 1,300 MWh, mainly to smooth wind-renewable resources. VRB Energy is already building a 100-MW, 5-hour storage facility in Hubei (see Figure 6.44), while Rongke is building a 100-MW, 4-hour system in 10-MW building blocks. In addition, several sites in California will be installed in the next couple of years, totaling over a GWh from several different manufacturers.

Figure 6.44 VRB Energy's energy storage installed indoors in Hubei, China. (Source: VRB Energy. Reprinted with permission.)

6.3.8 Zinc-Bromine

Another flow battery with a modest number of trial systems installed; all under 1 MW, is zinc-bromine (ZBFB). The promise of low cost is the biggest appeal of this technology.

6.3.8.1 Chemistry

Traditional ZBFB batteries are like VRFB in that there are two tanks, a stack with membranes, and a set of pumps and pipes that move the electrolyte around as power is needed. ZBFB batteries are different from VRFB in that, instead of metal ions changing ionic state and staying in solution, zinc metal comes out of solution and is deposited on a plate as the battery is being charged. Because of this, they are sometimes referred to as hybrid flow batteries. When the battery is discharged, zinc atoms are stripped of two electrons and go back into solution as an ion. In the other liquid loop, bromine Br_2 is converted to cations $2Br^+$ and goes into solution. The amount of energy storage is limited by two factors. One is the amount of electrolyte in the tanks. The other, unlike VRFB, is the allowed thickness of the zinc plating on the negative plate. If the zinc plating is too thick, it will bridge over to the positive plate and short out the cell. Because of this, zinc bromine flow batteries are designed with and for a specific power to energy ratio.

Negative electrode reactions:

$$Zn \rightleftharpoons Zn^{2+} + 2e \text{ (Discharged)} \tag{6.12}$$

Positive electrode reactions:

$$Br_2 + 2e \rightleftharpoons 2Br^- \text{ (Discharged)} \tag{6.13}$$

6.3.8.2 Cell Design

Like VRFB, traditional ZBFBs have two reservoirs, two loops of electrolyte, and a stack with a membrane and electrodes, shown schematically in Figure 6.45.

Primus Power had developed a new design that uses only one tank, one liquid loop, and no membrane (see Figure 6.46). They claim that their design is fundamentally cheaper, higher power, and more reliable than two-loop systems. The separation of liquid types is performed by gravity that keeps the bromide-laden liquid on the bottom of the tank and the zinc and bromine ion solution on the top. As the state of charge changes, the separation between the two liquids moves up or down in the tank. In fact, since the color of the liquids are different, it is very easy to see the state of charge by observing it through semitransparent tank walls.

Like a VRFB, the same electrolyte can be pumped through multiple series stack cells to achieve higher voltages. Primus Power focused on products that can be dropped in place and connected without much fieldwork.

6.3.8.3 Performance

The Primus Power product, which has the highest energy density of this chemistry, could get 250 kW and 1 MWh in a stackable 40-ft container. ZRFBs can scale their power and energy by adjusting the relative size

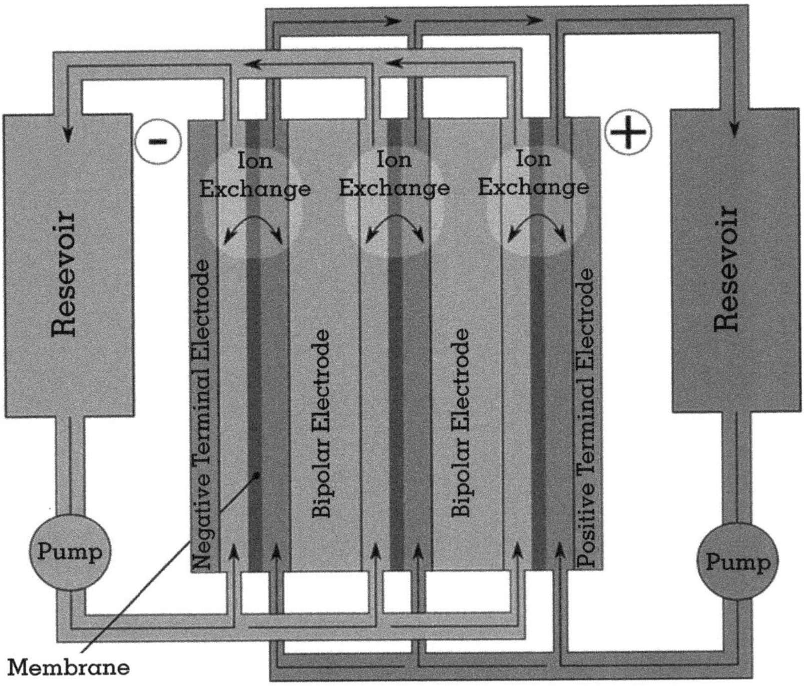

Figure 6.45 Diagram of a two-tank zinc-bromine system, showing three series cells supplied by the same electrolyte loops.

of the reservoir with respect to the stack and by adjusting the distance between plates in the stack as well. Primus elected to size their product at four hours.

Round-trip efficiency of ZRFBs varies between 64% and 73%, depending on the rate of discharge. Round-trip efficiency is a function of the size of the electrodes, and thereby the cost of the stack. Again, this election is made by the manufacturer when they decide which performance aspects are most important for their products. The losses in a ZRFB battery include the operating power of the pumps, conduction losses, self-discharge, and chemical reaction losses. Auxiliary power for pumping and cooling during cycling consumes about 6% of the pack's energy during one full cycle.

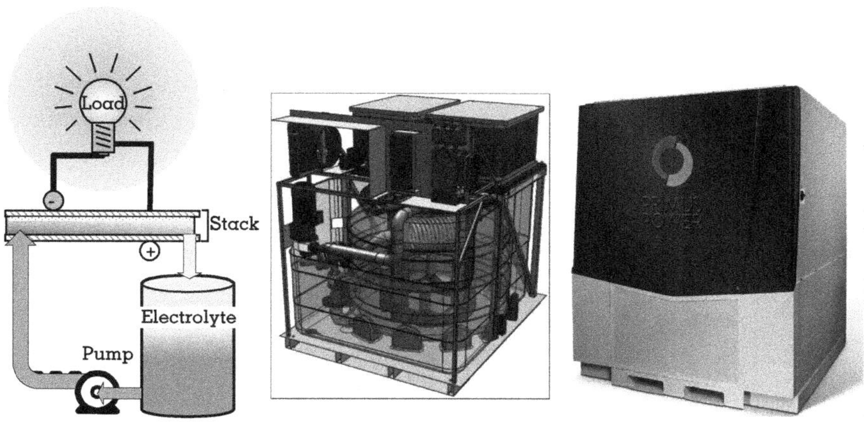

Figure 6.46 Diagram of Primus' one-tank zinc-bromine system and their product with and without skins. (Source: Primus Power. Reprinted with permission.)

Products using this chemistry should expect to be serviced every 7 to 10 years by replacing pumps and other moving parts. Designs with two liquid loops and a membrane will have to replace their stacks as well, since the membranes wear out with continued flow across them. The electrolyte, however, does not degrade and can be used indefinitely. Thus, the battery has an almost infinite cycle life but may need a serious overhaul at the end of 20 years due to the degradation of parts in the bromine environment.

The only caveat is that for every one to four complete cycles, a ZRFB battery must be completely discharged down to zero volts and held there for a few hours to completely remove the zinc from the electrode. Without this clean-out cycle, the zinc layer grows unevenly and could result in a shorted cell if tall lumps grow across to the other electrode. This maintenance cycle may be possible in applications where, on a daily basis, the energy storage is discharged until empty, then waits a period of time until it can be recharged. For example, it can power a nighttime load until empty, short its output with a shorting block, wait until the sun comes up, recharge slowly throughout the day, then repeat. Flexible power conversion technology would have to be implemented that can interface with a dc source at zero volts.

These products operate over a wide temperature range, but if continued use at hot ambient temperatures causes their electrolyte to exceed 40°C, a cooling loop through the tanks will be necessary. Unlike other batteries, the roundtrip efficiency peaks at 25°C operating temperature.

Like any other flow battery, there will be a few minutes delay between idle mode and full power, which includes the time to power up the pumps and get the electrolyte flowing in the stacks. If a quicker response time is required, pumps can be run continuously, but doing so will consume auxiliary power and reduce their service life.

6.3.8.4 Safety

The bromine solution in the electrolyte is caustic and will degrade many materials that it contacts. It is also dangerous to handle, and the fumes are toxic to breathe. These safety concerns are why manufacturers include double or triple walled protection again the tank leakage and try to avoid filling as a part of the installation process. Manufacturers are quick to point out that although their materials are hazardous, they are not nearly as dangerous as lead-acid, pool chemicals, and vanadium flow acids as depicted in Figure 6.47. The aqueous solution may be caustic, but it is not prone to thermal runaway, nor is it flammable.

6.3.8.5 Examples of Zinc-Bromine Energy Storage in Service

There are 19 ZBFB installed around the world, according to the GESDB [3]. The two largest were installed in Massachusetts, to perform renewable integration for wind and solar. Each could provide 500 kW for six hours as part of a smart-grid demonstration program at National Grid. As of this writing, one of them has been decommissioned for unknown reasons.

6.3.9 All Iron Flow Battery

ESS Inc., an energy storage systems provider using all-iron flow battery technology, just went public in 2021 and raised $465 million. Their technology uses iron dissolved in potassium-chloride saltwater, making it nontoxic, nonflammable, and low cost. Their product is built in modules that can be installed in containers or buildings and ships dry, so that it is easier to transport, and requires an addition of water to start working.

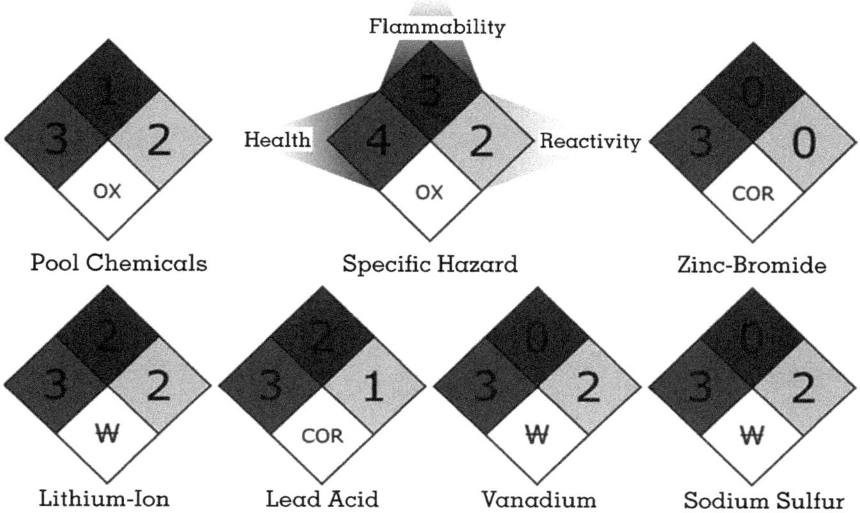

Figure 6.47 Hazards of zinc-bromine batteries vs standard swimming pool chemicals and other batteries.

6.3.9.1 Chemistry

Like a traditional flow battery design, iron flow batteries contain two reservoirs, two fluid loops, a stack with a membrane, and other balance of plant equipment as depicted in Figure 6.48. In the negative loop, dissolved ferrous ions (Fe^{2+}) gain two electrons during charge and become iron metal that plate onto the negative electrode. In the positive loop, dissolved ferrous ions lose an electron to form ferric ions (Fe^{3+}). Positive ions move across the ionically selective membrane to balance the electrical charge.

Negative electrode (anode) reactions:

$$Fe \rightleftharpoons Fe^{2+} + 2e^{-} \text{ (Discharged)} \tag{6.14}$$

Positive electrode (cathode) reactions:

$$2Fe^{3+} + 2e^{-} \rightleftharpoons 2Fe^{2+} \text{ (Discharged)} \tag{6.15}$$

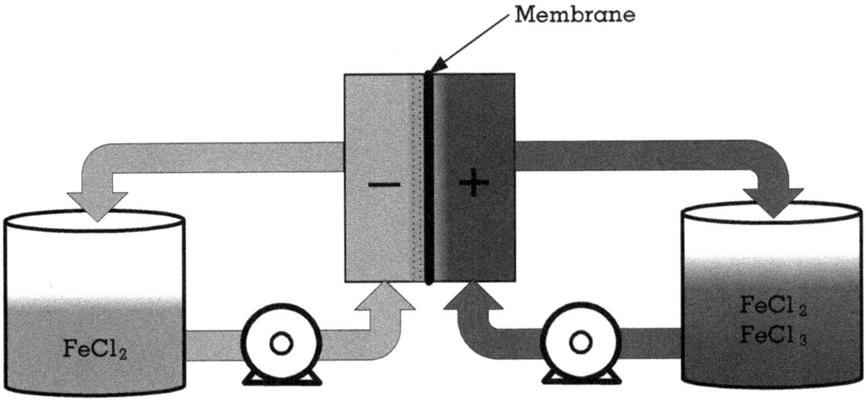

Figure 6.48 Two-tank, all-iron, hybrid flow battery.

6.3.9.2 Performance

ESS Inc. has a grid-ready energy storage product built into a 40-ft container format shown in Figure 6.49, that can provide 50 kW of continuous power for 12 hours or 90 kW for 4 hours. Its round-trip efficiency is between 70% and 75% at the dc terminals. For a complete ac to ac efficiency, detract another 4% to 6% round-trip for PCS, wiring, and trans-

Figure 6.49 Artist's rendition of ESS Inc. container-based energy storage system in service. (Source: ESS Inc. Reprinted with permission.)

former losses. The container, weighing 16,000 kg, ships dry so it can be transported without violating any intermodal transport weight limits. Another advantage of shipping dry is that the system is completely inert and has no stored energy hazards.

ESS Inc. claims that their system can achieve over 20,000 cycles and guarantees their products for 10 years but claims that they can last 25 with associated annual maintenance. Like the vanadium and zinc-bromine flow batteries, all iron flow battery electrolytes do not degrade with cycles, and the only things that wear out are the moving parts, such as pump impellers. Unlike the other flow chemistries, the electrolyte does not corrode internal components, so all iron flow battery systems stand a good chance to last its expected lifetimes.

6.3.9.3 Safety

All iron flow batteries use a safe, nontoxic, and even environmentally benign electrolyte. Not only is it nonflammable, but there is no chance for thermal runaway. However, as with any aqueous battery, if the terminal voltage exceeds that which can electrolyze water, there will be a danger of hydrogen evolution. This generation will only happen if all the ferrous ions are used up and plated as iron and the voltage spikes with continued, uncontrolled charging.

6.3.10 Zinc Batteries

In addition to zinc-bromine batteries, there are other nonflow, zinc-based battery technologies that are worth mentioning for grid-storage because of recent progress in their development. One of them is a zinc-alkaline battery developed by Urban Electric Power, NY, similar to the common nonrechargeable alkaline consumer batteries but designed to be recharged thousands of times.

6.3.11 Zinc Manganese Dioxide

Urban Electric Power (UEP) has figured out how to make the ordinary alkaline primary battery into a secondary battery. Earlier attempts at this challenge by others resulted in a consumer product that had a cycle life of anywhere from 10 to 100 cycles. UEP figured out how to make their battery cycle thousands of times and have a similar energy density as that of the mid-range lithium-ion cells. Additional work is being done at

Northeastern University and the City College of New York to increase the cycle life even further by adding various dopants such as bismuth and copper to the cathode.

6.3.11.1 Performance

Their 7.1m by 2m, 41,000 kg product shown in in Figure 6.50, contains 3.75 MWh of ac energy (750 kW for 5 hours) and comes complete with a dc/ac PCS, so it is ready to be connected to the grid through a step-up transformer.

Its round-trip efficiency is 86% ac/ac at the 5-hour rate, and its self-discharge is almost as good as lithium-ion at 20% per year.

Because it uses an aqueous electrolyte, ZMD cells do not go into thermal runaway even when subjected to the harsh UL 9540A test regimen, nail penetration, short circuits, overcharge, and over-discharge abuse testing.

6.3.12 Sodium-Ion Batteries

Sodium-ion batteries work similarly to lithium-ion batteries in that sodium ions are intercalated into anode and cathode materials. As of this writing, several companies are working on this technology, achieving various performance levels. The appeal of this technology is the lower cost of sodium compared to lithium since it is a more abundant natural resource than lithium.

6.3.12.1 Chemistry

Sodium ions are shuttled between the anode and cathode, typically through a nonaqueous electrolyte. The anode material is usually hard

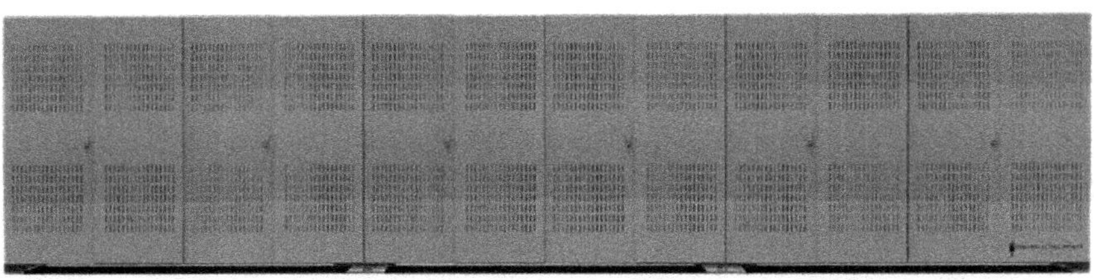

Figure 6.50 Urban Electric's custom enclosure can deliver 3.75 MWh of ac energy. (Source: Urban Electric Power. Reprinted with permission.)

carbon, although one company, Natron, is using a material called Prussian blue, an ingredient in the dye used to color blue jeans (Figure 6.51). Hard carbon has a randomized orientated structure that allows storage of the large sodium ion, whereas graphite, used by lithium-ion batteries, does not accept them. The cathode materials are either an oxide, Prussian Blue, Prussian White, or some proprietary mix.

6.3.12.2 Performance

Natron's Prussian Blue material is a large, stable matrix that readily allows sodium intercalation without flexing the structure. Although the Prussian Blue structure shown in Figure 6.51, is in both anode and cathode, they each have different corner elements, which will provide the cell's electromotive force. This design enables their battery's extremely high power-to-energy ratios (up to 60X) and cycle life (up to 25K cycles). However, the size of that structure also causes the cell's low volumetric density of 25 Wh per liter. Because of its high power and low self-discharge rate, Natron has been targeting the UPS market for short-duration, bridge-to-backup applications with the module shown in Figure 6.52.

Another sodium-ion manufacturer, Faradion uses hard carbon anodes and an oxide-based cathode, which allows its product to have energy densities in the range of some of the LFP lithium-ion cells (around 300 Wh/l). The downside to their design is that the cell only cycles hundreds of cycles. Because of their low cycle-life, their products have also been

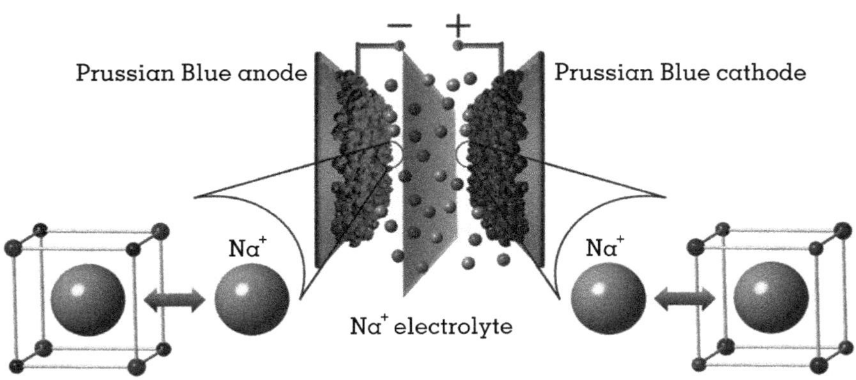

Figure 6.51 Natron's Prussian Blue, sodium-ion chemistry. (Courtesy: Natron Energy, Inc. Reprinted with permission.)

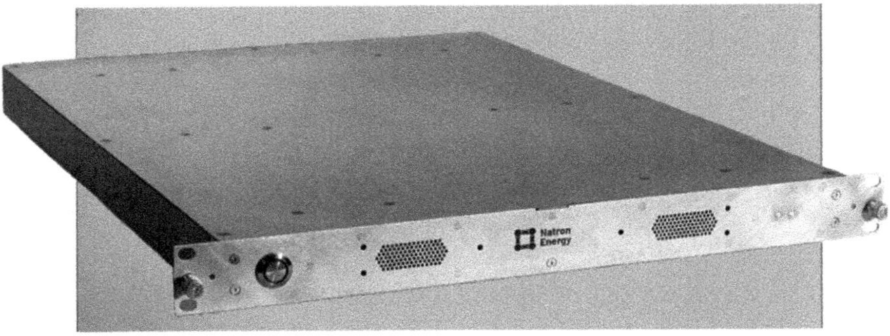

Figure 6.52 Natron's sodium-ion rack-mountable dc module can deliver 4 kW for 2 minutes, recharge in 8 minutes, and cycle 25,000 times. (Courtesy: Natron Energy, Inc. Reprinted with permission.)

targeted for back-up applications but for where power failures are rare, and cycle-life is not as important.

CATL is a relative newcomer in this technology, announcing in 2021 that they have a new sodium-ion cell with 160 Wh/kg (similar to LFP) using a hard carbon anode and Prussian white cathode. There are no specifics available, but CATL claims high cycle-life, 15-minute charging, lower temperature operation, and lower costs in their announcement [25].

Aquion was another sodium ion company that developed an aqueous electrolyte (as opposed to the typical nonaqueous type) for their battery. Its cathode was manganese oxide and its anode carbon. Sodium ions traveled through the sodium-sulfur saltwater to intercalate in the carbon or the MnO_2. Its product's attractive qualities were its abundant earth materials, nontoxic electrolyte, simplicity, and potential low cost. However, it had a low energy density of 17 Wh/l at a discharge rate of 4 hours, round-trip efficiency of 70%, and cycle life of 2,000–3,000 cycles.

6.3.12.3 Examples of Sodium-Ion Energy Storage in Service

Aquion had installed dozens of demonstration sites around the world and a handful of commercially active ones. The largest one on record is the 250-kW/1.25-MWh energy storage system that powers a 16-MW PV plant's auxiliary load during the night.

6.3.13 Liquid Metal Batteries

Ambri, a company in Massachusetts, is developing liquid-metal electrochemistry in which one or more of its active metal materials is liquid. In the labs of MIT in the early 2000s, Professor Donald Sadoway and graduate student David Bradwell researched dozens of different battery chemistries before settling on a liquid metal battery as the one with the most potential to succeed. As Sadoway is famous for saying:

> If you want to make a battery dirt-cheap, make it out of dirt.

The point is that the materials they chose to use in the cells are commonly available, contributing to the low expected cost of this technology. With venture capital, Sadoway, Bradwell, and Luis Ortiz started Ambri in the heart of Cambridge, Massachusetts.

6.3.13.1 Chemistry

Initially, Ambri chose two metals that would stay liquid at the elevated operating temperatures, separated by a layer of molten salt, as shown in the concept prototype in Figure 6.53. A simple explanation is that the density of the top layer is lower than that of the salt and the bottom metal, so it floats to the top when the cell is fully charged. When discharging, the top metal atoms lose electrons, become cations, and

Figure 6.53 Cut-away view of the inside of a conceptualized liquid-metal cell. (Provided by Ambri Inc. Reprinted with permission.)

move through the salt to alloy with the bottom metal atoms. The cell is completely discharged when the top layer is depleted. When charging, electrons are forced into the top electrode, where they attract the metal ions back up to the top again. Many other design details are necessary to make this work right and were the focus of Ambri's last decade of engineering effort.

Their latest cell design includes a bottom layer of antimony metal particles that do not melt at the cell's operating temperature. This proves to be an advantage because there is more surface area at which the top layer atoms alloy, thus the internal cell resistance is lower than with the previous metal choices.

The top layer metal is molten calcium, with a small amount of alloy to keep it molten at the 515°C operating temperature. It is also 100% compatible with the calcium-chloride molten salt through which its ions traverse to and from the bottom metal.

There are no side reactions in the cell during the charge and discharge cycling. Every electron that goes into the cell causes a calcium ion to move across the electrolyte. If the cell is overcharged or overdischarged some unintended side-reactions can occur. These reactions are not a safety concern, but over time, it can degrade battery performance. When operated within its specified SOC range and temperatures, there are no known degradation mechanisms in this chemistry. Oxygen and any impurities introduced during the production process are all reacted and immobilized during the first few hours of operation at high temperatures and do not impede the cell's operation into the future.

6.3.13.2 Cell Design

Everything about the cell is designed to withstand the 515°C operating temperature for the expected 20-year service life without degradation. The cell walls and internal support structures are all stainless steel. The negative terminal is insulated from the positive electrode cell case by a ceramic assembly that took years of development to achieve the targeted lifetime. Although conceptualized as a gravity-influenced stratification of molten layers, several designs have been developed using other constraining methods to facilitate a more flexible physical design for a more optimal system design. Ambri's latest cell design effectively flips the electrodes on their side, allowing the cells to have multiple vertically aligned negative and positive electrode plates, thereby allowing a significant increase in its capacity.

6.3.13.3 Performance

Ambri is still working on the final design, so what is presented will be a view of their current capabilities and potential for improvements. Each cell, 216 × 137 × 150 mm, can provide 400 Ah of capacity at 0.85 average voltage. Their target final production capacity will be 800 Ah, and they have a road map to get to 1000 Ah. The cells now have 50-Wh/kg and 75-Wh/l energy densities. If they achieve their energy targets, it may be possible to fit between 0.8 and 1 MWh into a 10-ft shipping container, including the requisite insulation. When such a container is fully outfitted with cells, conductors, and insulation, it will weigh approximately 50,000 lbs.

Their target dc to dc round-trip efficiency is 78% to 82% at CP/4 to CP/8, respectively. Engineering samples showed round-trip efficiencies as high as 87% and 91% at CP/4 and CP/8, respectively. Ambri states this was a result of the engineering prototype's specific design. Production cells will be lower than that for practical reasons. This is another example of making trade-offs in the product design for one performance factor over another. The design can be tweaked to improve efficiency, but other design aspects may be compromised, such as cost, energy density, and self-discharge.

The self-discharge mechanism is simply calcium ions occasionally bumping into antimony particles in the hot salt bath and inadvertently bonding. As with any battery, or chemical reaction for that matter, this process will increase with temperature, but at the nominal operating temperature of 515°C, the self-discharge rate is about 1% to 2% per day.

The cycle-life of this technology is this chemistry's strongest advantage. Ambri projects just 5% degradation over a 20-year service life with one cycle per day. As of yet, no known mechanism degrades capacity from cycling the cells. The biggest challenge for high-temperature batteries, like Ambri's, is temperature cycling. When the cells are heated to 515°C, cooled to ambient temperature, and reheated, tremendous amounts of expansion and resulting stresses occur in the cells, seals, connectors, bus bars, and structure. Ambri designed their cells to withstand at least one thermal cycle a year for 20 years. In reality, if the system needs no maintenance, there is no reason to thermally cycle the product on a yearly basis, so their product's service could very well be much longer.

6.3.13.4 Safety

The chemistry of liquid metal batteries is one of the safer ones out there, but there are a few cautions that cannot be ignored. On the positive side, the chemical reactions do not support any thermal runaway. If the temperature is increased enough, it may cause the steel cells to warp, but no thermal runaway will occur inside the cells. The heat generated by a short-circuit is not enough to overheat the cell, although it may accelerate aging by an incremental amount.

The very high temperature required to operate liquid metal batteries may be concerning to some and should not be ignored. The design of the Ambri enclosure keeps the external surfaces at a comfortable temperature, so it is safe to handle when operating. However, it is vitally important that the enclosure is built in such a way to prevent anyone from inadvertently accessing the inside of the enclosure until it has a chance to cool down. The structure and insulation thickness add to the physical protection around the cells, thus keeping them from coming in contact with the outside.

The calcium in the cell is considered to be a hazard by transportation authorities, because if it came in contact with water it would burn brightly. However, the enclosure, thick insulation, and thick stainless steel cell walls offer multilayered protection against exposure of this metal. In addition, when the cells are cooled below their operating temperature, they have no voltage and are essentially a solid block of metal and salts, with no possibility of having an electrical fault enroute. Nonetheless, since calcium is inside the cells, the product needs to be classified as a hazardous material for shipping purposes.

6.3.13.5 System Design

The cells must be kept at 515°C +/– 25°C to operate at their optimum performance potential. A thick insulating blanket has two main purposes. First as mentioned already, its purpose is to keep the outside surfaces at a touchable temperature. Second, it can retain the heat of the cycle losses so that auxiliary power is not necessary to maintain the internal temperatures. This requires a thick layer of highly thermally resistive material between the cells and the outside structure. It also requires careful mechanical design because many structural and electrical components link the cells and the outer shell, paying careful attention to their thermal conductivity and expansion coefficients. One of Ambri's prototype systems is shown in Figure 6.54.

Like lithium-ion cells, series-connected liquid-metal cells must be kept in balance using an external circuit when their SOCs diverge. This requires battery management at each string, dedicated to its optimum performance and protection. The cell voltage varies from 0.6V at the bottom of discharge to 1.2V at the top of charge. This wide range will require power conversion with a wider range than needed for other chemistries.

6.3.14 Nickel-Iron

Nickel-iron was invented in Europe before 1900 and was introduced in 1908 by Thomas Edison to the U.S. automotive market. He patented several improvements over the original (Figure 6.55) and tried them in his Pearl Street Station microgrid. Although the EV market fizzled and

Figure 6.54 Ambri's prototype containerized system. Inside, the cells are at 515°C, outside the container is lukewarm to touch (Provided by Ambri Inc. Reprinted with permission.)

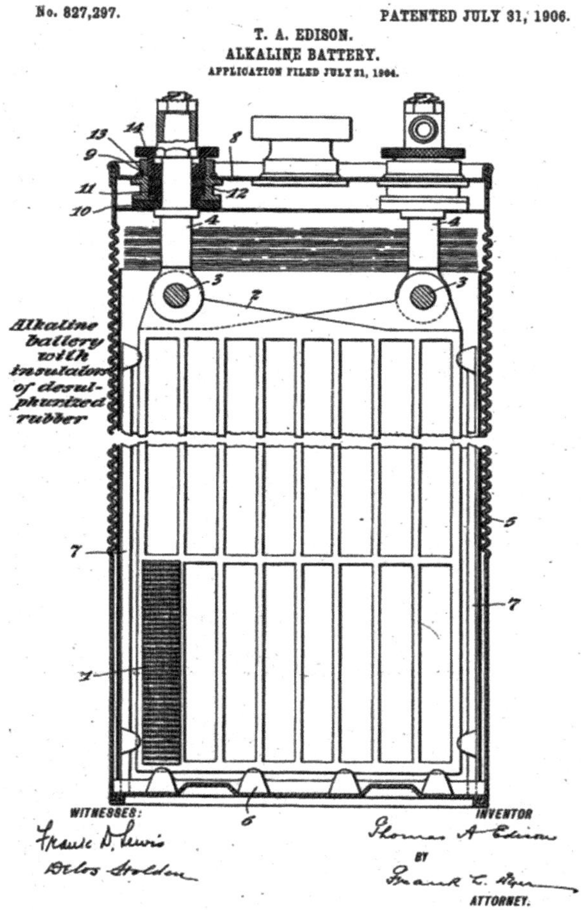

Figure 6.55 Thomas Edison's nickel-iron stationary battery. (Source: USPTO.)

died early in the twentieth-century, nickel-iron batteries remained a viable product for industrial traction applications because of their superior cycle life and abuse tolerance over lead-acid batteries.

Its characteristics, up until recently, would have made it a poor choice for stationary energy storage applications because of the battery's high self-discharge (24% per month), mediocre cycle life (< 2,000), and high cost of manufacture (complex processing steps).

Encell, a company started in 2009, improved nickel-iron battery performance to cycle over 11,000 times, float charge for up to 50 years, discharge up to 6C, have an energy density of 60 Wh/l, and self-discharge less than 10% per month. They also decreased the cost of manufacturing by simplifying the cell's construction. As a result, they claim industry-leading cost per discharged energy over its service life (more on that in the Chapter 7).

The cycle life coupled with low initial costs makes it a viable contender among other non-lithium technologies. However, its energy density and some other design aspects will affect its installation costs, so they need to be considered in the total cost of ownership assessment.

6.3.14.1 Chemistry

For an older technology, nickel-iron batteries have one of the most complex chemistries of modern electrochemical storage technologies [1]. There are multiple sets of chemical reactions and different side reactions that sum to the overall reaction of:

$$3Fe + 8NiOOH + 4H_9O \rightleftharpoons 8Ni(OH)_2 + Fe_3O_4 \text{ (Discharged)} \quad (6.16)$$

The shuttling ions are hydroxide (OH^-) while water goes in the opposite direction, so the net effect is an oxygen ion transfer from one electrode to the other during charge and discharge half-cycles. One point to note is that the electrolyte does not participate in the main cell reactions, which means it does not change its basic or acidic qualities as the cell changes SOC. Another characteristic is that when iron takes part in the reactions, its divalent ion bonds with one or more OH^- molecules, which is one of the factors that discourages iron from gathering and forming dendrites in solution.

Like lead-acid and other flooded nickel-based chemistries, cells of this chemistry will generate hydrogen and oxygen as it charges. During the first 80% of SOC, only 2% of the charge current goes toward hydrogen evolution, while in the last 20% of SOC, 25% of the charge current goes toward hydrogen evolution. At the top of charge, most of the current goes toward hydrogen evolution and the rest counters the 10% per month self-discharge current. This evolution of hydrogen means a loss of water, which must be replenished rather frequently (up to once per week in some applications).

6.3.14.2 Cell Design

Encell's cell design looks a lot like traditional nickel-iron cells, but inside is what makes all the difference. Encell calls it fused-iron technology because, instead of just crushing the electrode particles together and hoping that they make electrical contact with each other, they incorporate a formation cycle that fuses the particles at their touch points. This locks in their ability to conduct electrons to and from the reaction sites and dramatically improves the cell's power capability. The cells are made of a polymer case so they can be easily racked next to each other. They include a vent and a flame arrestor to allow the gasses to safely escape. In addition, they incorporate extra volume to allow for more electrolytes so that the frequency of water filling is not too burdensome on the operators.

6.3.14.3 System Design

Because of water loss, a system using these cells will probably include a self-watering system similar to that installed in the Alaskan Ni-Cd system. In addition, because of the hydrogen emission, constant air ventilation is required. Because they lose more energy near the top of charge, these cells will naturally self-balance with each other in series-connected strings, reducing the need for individual cell monitoring. Figure 6.56 shows a large difference between the upper charge voltage (1.75V) and the lowest discharge voltage (1.0V), therefore, a power conversion system with a wide input range would be required to fully charge and discharge the battery string.

Like other nickel-based chemistries, the nickel-iron cells have a peak voltage at the end of charge and may even go down if the charge current continues. Therefore, it is not recommended to connect cells or strings of cells in parallel. Each string would have to have its own power conversion to regulate it based on its SOC.

6.3.14.4 Examples of Nickel-Iron Energy Storage in Service

Encell has several behind-the-meter installations ranging from 60 kWh to almost 3 MWh. Figure 6.57 shows part of a 3 MWh behind-the-meter

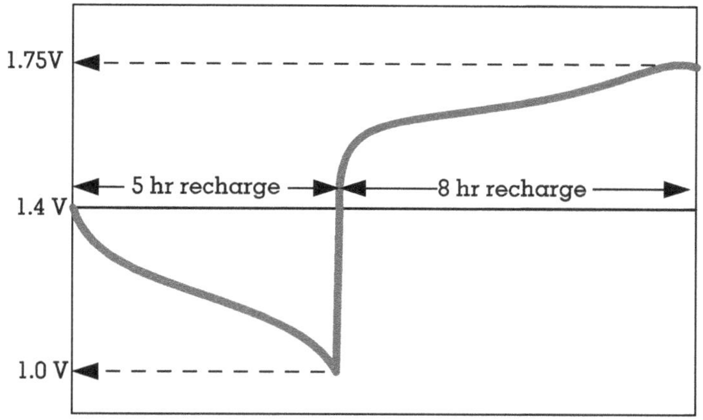

Figure 6.56 Nickel-iron's voltage range during discharge and charge cycles.

Figure 6.57 Examples of commercial (left) and residential (right) behind-the-meter energy storage using nickel-iron batteries. (Source: Encell Technology Inc. Reprinted with permission.)

installation in California, and one residential installation. Encell has one other large installation (1.9 MWh) in California, and few other commercial sites in New York City, the Caribbean, and others, as well as many residential installations in the United States and abroad. The main market, so far, has been demand-charge management, backup power, and solar integration.

6.3.15 Electrochemical Supercapacitors

The very first attempt to store energy used a Leyden jar that is essentially a capacitor. So, it would be appropriate to say that capacitive energy storage is the oldest form of electrical energy storage. Today, ordinary parallel-plate capacitors are used in almost every electronic circuit ever designed, manufactured, and dreamed of. Ranging from picofarads to millifarads, they serve a useful purpose of smoothing electrical power over the course of microseconds to tens of seconds.

In a standard capacitor (refer to Figure 6.58), two parallel plates are separated by a thin, insulating dielectric layer which allows difference in charge between the two plates to exist. The capacitance is:

$$C = \varepsilon A/d \tag{6.17}$$

Where ε is the dielectric constant of the material between the plates, A is the area of the plates, and d is the distance between the plates. Commercially available capacitors have many thin layers of metal films or foils and insulators to achieve the desired capacitance values. Common dielectric values for insulating materials range from two to 10 times that of vacuum, depending on the materials.

The voltage on the capacitor is proportional to the charge on each plate:

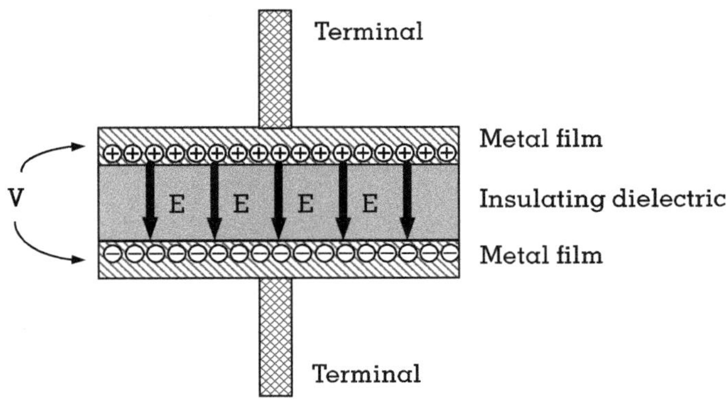

Figure 6.58 Standard capacitor consisting of two parallel plates separated by a dielectric layer.

$$V = Q/C \tag{6.18}$$

Where Q is the charge on one of the two plates. The voltage creates an electric field inside the capacitor, which is where the energy is stored. The energy in a capacitor is proportional to the square of the voltage on it:

$$E = \frac{1}{2}CV^2 \tag{6.19}$$

Therefore, a high voltage rating goes a long way toward its ability to store energy. A farad is a unit of capacitance that would correspond to two plates with an area of 22,000 football fields, separated by an airgap of just 1 mm. Yet even a 1F capacitor holding 100V would only contain about 1.4 watt-hours of energy.

In achieving higher energy capacities there are limits to how thin the layers can be and to how much voltage can be applied. If the insulators are too thin, high voltages will arc over and short the capacitor. Hence why most capacitors used in common electronic circuits are rated in units of μF, or one-millionth of a farad.

Supercapacitors are a class of capacitors that have more than a farad of capacitance, even thousands of farads in a single unit. Supercapacitors work by using an electrolytic fluid to increase the effective area of and decrease the effective distance between the plates. Each plate is made with a texturized surface, such as a porous material, like carbon particles, activated charcoal, and even carbon nanotubes. This increases the surface area of the plates by orders of magnitudes more than a smooth surface. An electrolytic fluid is placed between the plates, usually captured in a porous, ion-permeable film. When charge is transferred between the plates, the negative plate will attract positive ions in the electrolyte and the positive plate will attract the negative ions of the electrolytic pair. The ions stay in solution but are separated from the plate by the thickness of a molecule at all of its rough surfaces, as shown in Figure 6.59.

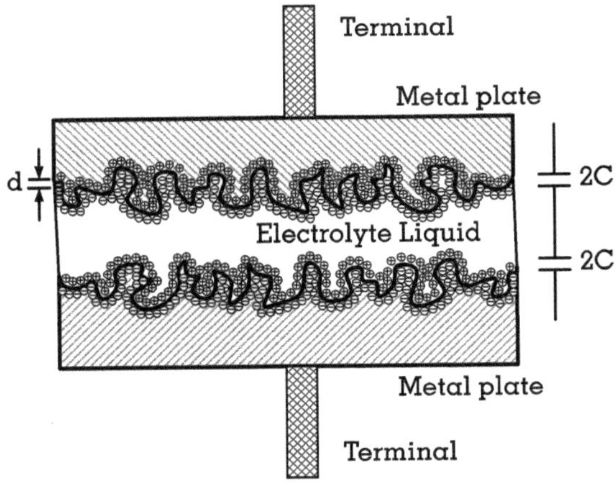

Figure 6.59 Electrolytic double-layer capacitor.

The charge build-up on one plate is mirrored in the opposite plate, causing the electrolyte to separate its charged ions. Because there are two layers in this structure, it is sometimes called an electrolytic double-layer capacitor. With the very large surface area compounded by the ultra-small distance between opposite charges, the capacitance can be extremely large. In addition, since no chemical reactions occur, the supercapacitor can be discharged rapidly and without any degradation to its capacity over millions of cycles. The energy density is much higher than ordinary capacitors, but not quite as high as electrochemical batteries.

Maxwell Technologies Ltd. of Korea and Skeleton of Germany offer products based on their 60138 cylindrical cells. Both manufacturers use symmetrical carbon-based electrodes that offer over 1 million full DoD cycles and 10 years of service life. Skeleton's graphene-powered cells have a capacitance of 5,000F, resulting in a practical energy density of 12 Wh/l when considering a reasonable operating range of 1.5V to 3V. Both offer full-scale energy storage systems serving short-term power needs such as UPS and grid stabilization applications.

A hybrid supercapacitor combines redox actions with double-layer action in one unit. This can increase the energy density but will affect its rate and cycle life abilities slightly. The general class of this hybrid approach is called a pseudocapacitor, where the ions transfer electrons with the plates in an intimate exchange at the plate surface. This provides a higher output voltage and larger energy storage per volume.

Taking the pseudocapacitor even further, a lithium-ion capacitor is a combination of lithium-ion anode with the faradaic cathode of a double-layer supercapacitor. Lithium ions intercalate into the negative plate (anode), receiving electrons in the process. Instead of ions simply accumulating at the surface as in a pseudocapacitor, many more ions can be stored inside the surface, thus increasing the energy density even further.

PolyJoule is developing this technology using a conductive polymer-based cathode at which negative ions gather when the cell is charged. Their cell construction is similar to that of a traditional battery or supercapacitor: anode, cathode, liquid electrolyte, and separator, rolled into a 50100 can as shown in Figure 6.60. Being part capacitor, it has a wide operating voltage range under which a meaningful charge is stored. Although, theoretically, the cell can withstand a range from 0V to 4.7V, PolyJoule recommends continuous cycling voltage range between 0.5V and 3V for their cell.

The polymer-carbon hybrid electrode technology has the advantage of long cycle life with a predictable decay curve and the absence of

Figure 6.60 A 2.2-Wh lithium-ion hybrid supercapacitor from PolyJoule. (Source: PolyJoule Inc. Reprinted with permission.)

an accelerated lifetime degradation past any EOL threshold. Currently, PolyJoule expects its cells to reach 80% initial capacity after 12,000 full DoD cycles, and continue to lose capacity as it cycles in a linear manner all the way to zero capacity.

Another advantage of this cell is its relative safety. The nonvolatile, nonflammable electrolyte, no lithium metal, and no metal oxides, guarantee no thermal runaway conditions, regardless of nail penetration, high temperature heating, overvoltage/overcurrent, or short circuit. The cell material combinations also allow for a relatively large operating temperature window (–40 to +40 °C). PolyJoule's 50100 cells cycle at a 1-hour rate and produce no noticeable heat, at a cell level, or even when enclosed in a pack.

Its high power to energy capability relegates it into applications of high-power, low-cost applications, or nonmotive, long-lifetime, low-cost storage applications where volume and weight are less of a constraint and lifetime safety remains important.

6.3.15.1 Examples of Supercapacitors in Service

According to the DOE's database of energy storage [3], over 30 MW of electrochemical supercapacitors have been installed around the world, ranging from 30 kW to 4 MW in size for each installation. Most of the installations are in Korea and are used to support the grid at subway stations. These 1- to 2.5-MW installations were built by Maxwell and are primarily used to absorb braking energy from the trains that enter the station. The trains inject energy into the power line as they decelerate to a stop. The extra power is absorbed by the supercapacitors and is then used to help accelerate the train out of the station. They claim that this process has helped to reduce power consumption by 20% overall, and it improves the power quality in and around the stations.

6.4 Summary of Electrochemical Energy Storage

This chapter described a dozen or so battery chemistries, their designs, and some examples of their use for grid applications. It showed how cell chemistry, design and overall system design can influence key performance factors of an electrochemical energy storage system. Cell design variations of some chemistries can give them a wide range of capabilities including cycle-life, and energy to power (e.g., changing the formulations of lithium-ion cells). Some chemistries can be inherently safe, while others present a challenge in that regard. In any case, even the safest batteries can be made dangerous with careless system designs, poor installation, and user errors. It is encouraging to see old battery chemistries become promising new ones when scientists and engineers get together to improve upon their construction. However, it is disappointing to see failures of companies who come up with bright ideas but could not keep the momentum going. In this century, there have been great strides in improving battery capabilities, service life, and costs, and, as a result, they will have a future in making renewable energy a big part of this planet's future.

Batteries have a wide range of efficiencies, costs, power, and energy, but the one thing that batteries have over any other energy storage medium is their power and/or energy density. This trait allows them to be placed where they are needed, whether it is a rack in a basement, a container behind a factory, a few containers in a switchyard, an urban building, enclosures sprinkled around a PV farm, or at the base of a wind turbine. Local placement enables their ability to provide lower-cost energy where it is needed and to help support the local grid's power quality. Batteries are also extremely flexible, in that they can be configured or designed to respond almost instantly, delivering extremely high power for short or long durations. This allows them to be used in a broad array of grid-supportive services.

The issue up until recently has been their costs. Whether they cost too much upfront, or they cost too much to maintain, their total cost of ownership exceeded that of other forms of energy storage. In the wake of recent EV developments, their prices and capabilities have improved dramatically, allowing them a foothold in grid-related markets.

Table 6.1 is a summary of the technologies discussed in this chapter. A thorough reference for batteries of all types is in [1].

6.5 Summary of Other Chemistries

Table 6.1 Summary of Cell Chemistries Described in This Chapter

Chemistry	Advantages	Challenges	Best Applications	% of Total Battery Installs as of 2020 (by Power)
Lead-Acid	Low cost; High recyclability; Moderate abuse tolerance; Self-balancing.	Low energy density; Low cyclability; Medium calendar life; Environmental hazard; Watering required.	UPS; Infrequent peak-shifting; Microgrid back-up; Spinning reserve.	< 5%
Nickel-Cadmium	Long calendar life; Moderate abuse tolerance; Self-balancing; High power.	Low energy density; High cost; Medium cyclability; Environmental hazard; Watering required.	UPS; Peak-shifting; Microgrid back-up; Spinning reserve.	< 1%
Nickel-Metal Hydride	Moderate abuse tolerance; Self-balancing; High-power.	Medium energy density; Medium cyclability; High cost.	UPS; Peak-shifting; Microgrid back-up; Spinning reserve.	< 0.1%
Lithium-Ion	High cycle life; Long calendar life; High energy density; Flexible design; High efficiency; Low self-discharge.	Low abuse tolerance; Moderate cost; Requires cell balancing/ monitoring; Thermal runaway potential.	Frequency regulation, ; Spinning reserve,; Peak shifting; Renewable integration, ; Capacity market; Arbitrage	< 80%
Sodium-Sulfur	High cycle life; Long calendar life; Medium energy density; Low self-discharge; Moderate cost; Hot ambient temperature tolerance	Low power; Low efficiency; Severe fire reaction to internal damage	Load shifting; Arbitrage; Capacity market; Renewable firming.	5%
Vanadium Redox Flow	High cycle life; Long calendar life; Flexible power and energy configurations; Reusable electrolyte at EOL; Nonflammable.	Low energy density; Low efficiency; High auxiliary load; High cost.	Load shifting; Capacity market; Renewable firming; Frequent peak shifting.	< 7%

Table 6.1 (continued)

Chemistry	Advantages	Challenges	Best Applications	% of Total Battery Installs as of 2020 (by Power)
Zinc-Bromine Flow	High cycle life; Long calendar life; Flexible power and energy configurations; Reusable electrolyte at EOL; Nonflammable.	Low energy density; Low efficiency; High auxiliary load; Moderate cost; Requires regular stripping cycles.	PV load shifting; Capacity market.	< 3%
All Iron Flow	High cycle life; Safe materials.	Low energy density; Low efficiency.	Load shifting; Capacity market; Renewable firming; Frequent peak shifting.	0.01%
Zinc Manganese Dioxide	Low cost; Medium energy density; Medium power; Nonflammable.	Medium cycle life.	UPS; Infrequent peak-shifting; Microgrid back-up; Spinning reserve.	0%
Sodium-Ion	Benign materials; Some have high cycle life; Some have high rate.	High cost; Some have low cycle life; Some have low rate; Low energy density; Requires cell balancing/ monitoring.	UPS (short duration).	< 0.1%
Liquid Metal	High cycle life; Long calendar life; Ambient temperature tolerance; Moderate energy density; Abuse tolerance.	High temperature; Unproven; Low power; Requires cell balancing/ monitoring.	Capacity market; Renewable firming; Frequency regulation; UPS; Microgrid backup; Spinning reserve; Frequent peak-shifting.	0%
Nickel-Iron	Low cost; High cycle life; Long calendar life; Abuse tolerance; Self-balancing.	Low energy density; High float-currents; Frequent watering required.	UPS; PV load-shifting; Peak-shifting; Microgrid back-up; Spinning reserve; Capacity market.	0%
Electrochemical Supercapacitors	High power; High cycle life; Abuse tolerant; No thermal-runaway.	Low energy density; Wide operating voltage; Some have high self-discharge.	UPS; Voltage stabilization; Impulse power; Frequency regulation.	< 1%

Table 6.2 Summary of Other Cell Chemistries Not Discussed in This Chapter*

Chemistry	Characteristics	Disadvantages
Lithium Metal	Solid lithium anode coupled with various cathodes similar to lithium ion.	Extreme reaction to abuse; Low cycle life; Restricted transportation options.
Lithium Iron-Sulfide	Similar in concept to the liquid metal battery; High temperature; Abuse tolerance; High energy density.	High cost; Low efficiency.
Nickel-Hydrogen	Long calendar life; Many short cycles; Pre-cursor to NiMH.	Expensive.
Nickel-Zinc	Lightweight electrodes.	Pretty good cycle life; Not good for long float charges; Prone to dendrites if not cycled continuously.
Silver-Zinc (Silver-Oxide)	Very high energy density.	Expensive; Poor cycle life.
Silver-Cadmium	High energy density.	Mediocre cycle life; Really expensive.
Sodium Nickel-Chloride	High temperature similar to sodium sulfur; Low power.	Mediocre cycle life; Expensive.
Zinc-Air	Oxygen electrode (air).	Short service life; Low efficiency.
Zinc-Carbon	Low cost; Moderate cycle life; Moderate calendar life; Abuse tolerance; Nonflammable.	Low efficiency; Low energy density; High self-discharge; Low power; Requires periodic stripping cycles.

*Characteristics as of 2021—some will change as new products are developed and evolve.

References

[1] Linden, D., *Handbook of Batteries Part 1: Principles of Operation*, McGraw-Hill Companies, 2004.

[2] Hoff, C., *Energy Storage Architecture (Elements in Grid Energy Storage)*, Cambridge: Cambridge University Press, 2022.

[3] Nguyen, D. T., "DOE Global Energy Storage Database," 2021.

[4] Rodriguez, G. D., W. Spindler, and D. S. Carr, "Operating the World's Largest Lead/Acid Battery Energy Storage System," *Journal of Power Sources*, 1990, p. 10.

[5] www.fireflyenergy.com.

[6] Firefly, "Firefly User's Manual for the 'OASIS' G31 & 4V/450AH," Firefly, August 2021, https://docs.google.com/document/d/1Caqy77A-sH8plNlibhJjDoaSwO3XPqQ89GKx-miYBG4/edit#.

[7] El-Nakeeb, M. A., A. E. El-Degwy, and E. E. Khalil, "Effect Double Vertical Batteries Storage on Hydrogen Concentration in Air-conditioned Storage Room," in *ISHVAC2017*, Jinan, China, 2017.

[8] Hoff, C., and K. Steeves, "New Insights into Thermal Runaway of Valve Regulated Lead-Acid Batteries," *Battcon*, 2005.

[9] Stolte, W. J., "Chino Battery Energy Storage Power Plant: First Year of Operation," EPRI, December, 1992.

[10] GVEA, "Battery Energy Storage System (BESS)," GVEA, 2019, https://gvea.com/battery-system/.

[11] Saft Batteries, "NiCd Block Battery Technical Manual," July 2004, https://us.v-cdn.net/6024911/uploads/attachments/379/4108.pdf.

[12] DeVries, T., "World's Most Powerful BESS is Online and Working in Alaska," Golden Valley Electric Association, Fairbanks, AK, 2005.

[13] ABB Inc., "Case Note: World's Largest Battery Energy Storage System," ABB Inc, Fairbanks, Alaska, USA, 2011.

[14] Ellis, T., "GVEA Considers Upgrading or Replacing BESS, the Aging, Outage-preventing Backup Battery," *KUAC*, April 8 2021, https://fm.kuac.org/energy-environment/2021-04-08/gvea-considers-upgrading-or-replacing-bess-the-aging-outage-preventing-backup-battery, https://fm.kuac.org/energy-environment/2021-04-08/gvea-considers-upgrading-or-replacing-bess-the-aging-outage-preventing-backup-battery.

[15] Jang, J., et al., "A Review of Functional Separators for Lithium Metal Battery Applications," *Mterials*, October 2020.

[16] Golubkov, A., et. al., "Thermal-Runaway Experiments on Consumer Li-ion Batteries with Metal-Oxide and Olivine-Type Cathodes," *RSC Advances*, Vol. 2014, No. 4, November 2013, pp. 3633–3642.

[17] IEA.org, "Energy Storage," IEA, Paris, 2021.

[18] Fortune Business Insights, "Battery Energy Storage, Global Market Analysis, Insights and Forecast 2017–2018," Fortune Business Insights, Maharashtra, India, 2021.

[19] Colthorpe, A., "Battery Storage as Peaking Capacity: How Alamitos Changed the Game for California," *Energy Storage News*, June 7, 2021.

[20] https://hornsdalepowerreserve.com.au/.

[21] Mackenzie, W., "U.S. Energy Storage Monitor," Wood Mackenzie, P&R/ESA , March 2021.

[22] BNEF, "Global Energy Storage Market Set to Hit One Terawatt-Hour by 2030," *BloombergNEF*, November 15, 2021.

[23] NREL, "Grid-Scale U.S. Storage Capacity Could Grow Five-Fold by 2050," National Renewable Energy Laboratory, 2021.

[24] Schenkman, B., and D. Bormeo, "Sandia Third-Party Witness Test of UniEnergy Technologies 1 MW/3.2 MWh Uni.System™," Sandia National Laboratories, June 2015.

[25] CATL, "CATL Unveils Its Latest Breakthrough Technology by Releasing Its First Generation of Sodium-ion Batteries," July 29, 2021, https://www.catl.com/en/news/665.html.

CHAPTER 7

The Value and Costs of Energy Storage

Energy storage has come a long way from when in 1745 Ewald Georg Von Kleist zapped himself with the first Leyden jar, to when the Swiss first tamed the Limmat River to keep their industry ticking in 1890, to now when batteries are enabling significant reductions of the use of fossil fuels on the roads and the grid. This book describes the many different types of energy storage that have been explored by the power industry, each with its own sets of advantages and disadvantages. For example, pumped hydro energy storage lasts a long time, but it occupies an enormous amount of space. Thermal energy has a very high volumetric energy density, but its round-trip efficiency is lower than most others. Electrochemical energy storage has wide ranges of energy density, power, energy, efficiency, and cost. Scientists are continuing their efforts to find the perfect combination of elements and materials to produce the ideal battery having high energy density, infinite cycle life, long calendar life, fast recharge times, and, of course, a low cost.

In the meantime, storage users want to know how they can assess the options that are available now so that they can decide which energy storage is the most valuable for their present and future applications. Additionally, storage technology providers want to know which product attributes are the most important, so they can concentrate on those that meet the needs of their future customers. This chapter will present some simple tools to compare the attributes relative to the value they enable from the storage.

7.1 Levelized Cost of Energy

Typically, power generating plants are evaluated based on a metric of levelized cost of energy (LCOE) which is a ratio of the power plant costs, present and future, over its delivered energy. It is a levelized metric because it calculates the net present value (NPV) of future costs and delivered energy using a project-specific discount rate.

$$LCOE = \frac{Capex_0 + \sum_{t=1}^{T} \frac{Opex_t}{(1+r)^t}}{\sum_{t=1}^{T} \frac{Energy_t}{(1+r)^t}} \tag{7.1}$$

Where $Capex_0$ is the initial investment, $Opex_t$ is the yearly operating costs (Including fuel), T is the total number of service years, t is the year corresponding to $Opex_t$, and r is the discount rate.

LCOE for power plants allows financial analysts to compare the economic viability of one power plant against another even though they each may have different life spans, up-front and ongoing cost structures, and risks. Implicit in this equation is that the power plant's value is the energy it produces. The value of energy storage, on the other hand, is different and merits a new equation.

Storage's value could be the revenue from buying and selling energy at strategic price differentials, stabilizing the grid with tactically timed power injections or withdrawals, keeping a factory humming when distributed power is scarce or nonexistent, or a multitude of other services, as described in Chapter 2. The financial revenue models for how storage is compensated are not only complex but vary with the type of service, the market, utility, and country. Therefore, it is difficult to rely on one cost metric for all applications. We can, however, identify two measurable value streams that storage provides and base the cost of storage with respect to each one in a meaningful way.

7.2 Levelized Cost of Storage

The first value of storage is its *ability* to provide a specified amount of power and energy. In certain applications, the dominant value of the energy storage is its ability to provide energy, not necessarily how many times it has delivered (or can deliver) it. Examples of such applications are backup power (UPS) and capacity markets, where the asset is compensated by its availability to provide power for a certain duration. Levelized cost of storage (LCOS) is calculated by dividing the total cost of ownership (TCO) by the guaranteed capacity throughout the service life. It accounts for all the costs of buying, installing, and maintaining an ESS and the effects of capacity loss throughout the service life.

$$LCOS = \frac{Capex_0 + \sum_{t=1}^{T} \frac{Opex_t}{(1+r)^t}}{Energy\_Capacity_{EOL}} \tag{7.2}$$

In some instances, the desired value is power_capacity. So, one might think that a valid denominator in those cases is the power capability of the energy storage. However, even when a PPA contract asks for a certain amount of power and compensates for its availability, there is always an underlying requirement for that power deliverable over a specified length of time. Thus, the service being provided is actually energy, even though it explicitly enumerates power. In summary, LCOS is the cost necessary to ensure that a certain amount of energy can be provided when it is needed.

7.3 Levelized Cost of Using Storage

The second value of energy storage is the energy that it actually delivers performing a grid-supporting service. Levelized cost of using storage (LCUS) is a metric tracking the cost per *delivered* energy. Storage can both deliver and absorb energy and derive value from both directions. A proxy for both is simply the discharged energy. Like the LCOE and LCOS metrics, the numerator is the TCO, but LCUS' denominator is the discharged energy at the point of interconnect (POI). LCUS accounts for all the costs of buying, installing, charging, and maintaining an ESS, and the effects of wear and tear from its cycling with respect to a total amount of discharged energy. Since the value of the discharged energy is often monetized in proportion to the discharged energy, the sum of future energy discharges is calculated on an NPV basis as in the LCOE metric for power plants.

$$LCUS = \frac{Capex_0 + \sum_{t=1}^{T} \frac{Opex_t}{(1+r)^t}}{\sum_{t=1}^{T} \frac{Energy\_discharged_t}{(1+r)^t}} \tag{7.3}$$

This metric is quite useful in comparing ESS used in cyclic applications, where its value is proportional to the energy it cycles. In summary, LCUS is the cost necessary to deliver a certain amount of energy to the grid it serves.

7.3.1 Up-Front Capex

$$LCUS = \frac{\boxed{Capex_0} + \sum_{t=1}^{T} \frac{Opex_t}{(1+r)^t}}{\sum_{t=1}^{T} \frac{Energy\_discharged_t}{(1+r)^t}}$$

The up-front cost of the energy storage includes not only the equipment, but other expenses such as the installation, land rights, permitting, studies, transportation, engineering, and construction (see Table 7.1). Each technology has its own set of costs associated with installation that must be considered for a thorough comparison. As an extreme example, water is virtually free, but the reservoir in which it is stored is not. Air is free, but the machines used to compress it into a cavern are not. In the other extreme, energy-dense batteries are expensive, but their installation is cheaper than that of a larger-format battery with the same energy capacity, and certainly less expensive than carving out the middle of a mountain to accommodate a pumped hydro station.

7.3.1.1 Equipment

Energy storage equipment is rated to handle the site's power and energy requirements. Some equipment is rated solely on power, such as transformers, PCS, and wiring. Others have limitations of both power and energy. Batteries, for example, provide energy capacity but are limited

Table 7.1 List of Some of the Main Up-Front Capex Cost Items

	Pumped Hydro	Compressed Air	Gravity	Flywheel	Electrochemical
Equipment	Pumps/ turbine, motor/gen, transformers, switchgear, aux power, controls, flow gates, sensors, meters.	Compressors/ turbines, filters, thermal storage, furnace, motor/gen transformers, switchgear aux power, controls, valves, sensors, meters.	Motor/gen, cables, pulleys, mass, structure, controls, transformer, switchgear, aux power.	Flywheels, silos, power converters, transformers, switchgear, aux power, controls.	Batteries (including controls, protection, fire suppression, cooling, heating, and environmental protection), power converters, transformers, switchgear, aux power, controls.
Preparation	Surveys: geological, seismic, ecological, public use, environmental, safety, grid benefits, transmission, meteorological, archeological. Permits: land-use, water-use, interconnection, corps, dam, waterways, electrical, building.	Surveys: geological, seismic, ecological, environmental, safety, noise effects, grid benefits, transmission, archeological. Permits: land-use, interconnection, electrical, building.	Surveys: geological, seismic, safety, light pollution, grid benefits, transmission, noise effects. Permits: land-use, interconnection, electrical, building.	Surveys: geological, safety, noise effects, grid benefits, transmission. Permits: interconnection, electrical, building.	Surveys: geological, fire-risk, ecological, safety, arc-flash, seismic, grid benefits, transmission. Permits: fire protection, interconnection, electrical, building.
Installation	Clearing, excavation, blasting, cavern, sealing, dam(s), penstocks, diversions, power station construction, equipment installation, wiring, testing, commissioning, interconnection.	Cavern hollowing and sealing, land prep., power station construction, equipment installation, piping, wiring, testing, commissioning, interconnection.	Foundations, excavation, construction, assembly, mass construction, wiring, testing, commissioning, interconnection.	Site-prep, foundations, equipment installation, wiring, testing, commissioning, interconnection.	Site-prep, foundations, equipment installation, wiring, testing, commissioning, interconnection.

to a specified power rating, so a battery with a 2-hour rate capability may not be able to deliver all its energy in just 1 hour. A 100-MW/200-MWh battery is a two-hour system. Two of these batteries are required to provide 200 MW, even though together they add up to twice the required energy. In other technologies, such as pumped hydro, all their equipment is sized for power capacity, not energy. Additional energy is added by making the reservoir larger. Compressed air is the same. Its compressors, electric machines, pipes, and switchgear are all sized for power. This is important to keep in mind because the cost per energy of two different technologies that have different power-to-energy ratios will be distributed differently over two energy capacity bases. For example, a 1-MW power converter on a 10-hour system will be much cheaper than the same 1 MW power converter on a 2-hour system, with respect to the site's energy capacity. In another example, VRFB, like PHS and CAES, can be independently sized for power or energy. The stacks, pumps, and plumbing are sized for a site power rating, while the tanks of electrolyte can be independently sized for energy capacity.

Equipment versus Construction Materials

With some kinds of energy storage, there is a grey line between equipment and construction materials. Often, pipes, wires, ducts, building materials, are classified as construction materials. This may cause a shift between the costs under equipment and installation. For example, in a CAES design, there are multiple sets of pipes, valves, instruments, pumps, and other plumbing-type materials that would have to be installed on a CAES facility. Modular CAES designs [1] would have much of this hardware pre-installed in a factory-built box and would shift not only the construction materials costs into the equipment bucket, but the installation labor as well. Many battery designs are moving in this direction. Whereas a few years ago batteries were partly assembled at the job site, now many products are almost completely factory-assembled, saving time and labor at the installation site.

Oversizing for Life

Some equipment providing energy capacity, notably batteries, will degrade with time and use. If the application requires a certain amount of energy throughout the service life, then this will have to be accounted for in the up-front and operating costs. A brute-force method to provide a contracted capacity throughout the life of the battery is to oversize it at the start (see Figure 7.1). The advantage is that the construction, assem-

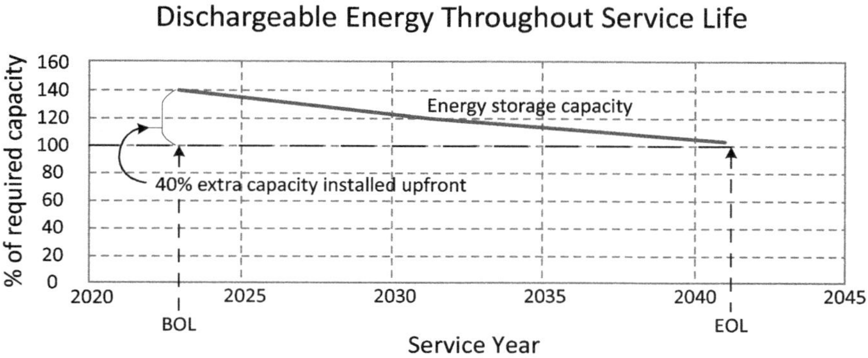

Figure 7.1 Oversizing an energy storage system's energy capacity by 40% at the beginning of life is one way to ensure its capacity remains above its requirement throughout its service life.

bly, shipping, procurement, installation, testing, and commissioning all happen only once, at the beginning. One disadvantage is that the batteries will start their calendar degradation earlier, thus accelerating the total degradation over life and requiring more energy storage to be installed as opposed to a more staged approach.

Augmentation

A second approach is to augment the installation's energy capacity as it degrades. The strategy would be to install a slightly higher energy capacity than required at the start. Then, after a few years, add additional capacity and repeat as necessary throughout the service life (see Figure 7.2). This option allows for a lower up-front equipment cost, spreads the cost of capacity degradation over the whole service life as a service-cost, and allows for some flexibility if the usage or performance is not as predicted. A speculative benefit is that battery prices will continue to decrease as they have in the past. This year, 2022 was an exception, but for the last 10 years, battery prices have been dropping over 15% per year. If this trend continues, even at a less aggressive pace, the cost of augmentation in the future will be lower than it is today. Additionally, battery technology has been improving. Therefore, chances are, when an operator seeks to add energy capacity to the installation, the available options will be improved, making it easier to conduct the augmentation.

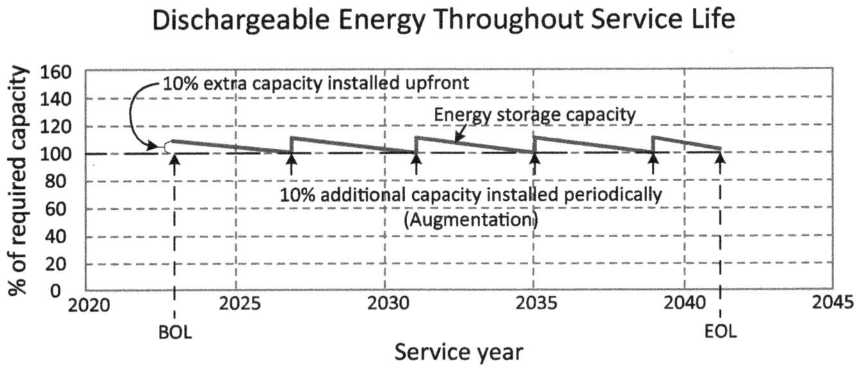

Figure 7.2 Another way to maintain storage capacity over a service life is to periodically augment the original installation with additional storage, as shown here every 4 years.

Either approach requires a deep understanding of the product's performance and its expected use, both of which are sometimes hard to nail down. First, the battery's life performance is usually predicted from a limited set of data over a short timeframe because the speed of recent battery development guarantees there is no time for a full 20-year service test before the next product evolution is released for sale. Instead, vendors rely on extrapolations based on their experience and the trends seen in accelerated test data.

Second, customer use cases can change from year to year. Current market conditions have the biggest influence on the way customers use energy storage, and both can change dramatically during the service life. For example, the PJM ISO in the northeast United States started with one profile for its frequency regulation assets to follow, and 3 years later changed its protocol to one that essentially demanded three times as much cycling in a given period. For some battery-based energy storage assets, this ruined their economic viability because their degradation rate accelerated more than originally anticipated.

Some storage operators, notably conservative utilities, will take care of their assets and use them gently only when they need to stabilize their grid, trying to make their rate-payers' investment last as long as possible. Other operators, notably IPPs, make money from the discharged energy and will run their batteries like a mustang on a racetrack. As long as it nets a profit during its service life, they use it to its max. Finding a medium between the two is possible, but requires analysis and patience.

Balance of Plant Equipment

Equipment costs also include those which do not directly provide power or deliver energy. These include enclosures, buildings, cooling, heating, auxiliary power, controls, maintenance tools, spares, personnel facilities, safety equipment, fire protection, explosion mitigations, personnel protection, site security, beautification, and even visitor accommodations. These items can be quite extensive depending on the technologies involved and should not be underestimated. Particularly for battery energy storage, these items are becoming more and more important as the price of batteries are becoming lower and lower. This year, industry trackers are saying the cost of battery packs has reached its all-time low of $120 per kWh. To make the rest of the energy storage system, the balance of plant must be considered. How much is the building? How much is the air-conditioning, the auxiliary power transformers, and so on? Those items do not go down in cost as the battery prices go down. Some of them even go up.

Transportation

Not to be ignored, transportation costs can be significant for the project installation. Grid-sized energy storage projects will invariably involve moving large equipment or a lot of smaller equipment. This can amount to a noticeable percentage of the project cost. With batteries in particular, large numbers of small- to medium-sized equipment will be transported, and many of these shipments are classified internationally as hazardous goods. This classification requires special packaging, limits transportation options, and increases handling fees. If some of the shipments are overweight by international standards, they will be more costly to move across borders and on roadways. Alternatively, manufacturers may transport them in multiple shipments and assemble them on site.

7.3.1.2 Site Preparation

Every storage project involves risks, risk of technology efficacy, market viability, product performance, installation troubles, environmental impact, earthquakes, floods, fires, and even the neighbor's complaints. Those working in this field have seen unexpected technology issues causing wholesale recalls, markets switching midlife to more extreme profiles, unexpected product performance—both good and bad, poor installation practices, unexpected water tables, spotted amphibians stopping bulldozers in their tracks, 300-year floods, and neighbors complaining of site-lines and noise factors.

Proactively conducting studies ahead of time will limit the impact these risks have on a project. Sometimes, they are mandated by the local authority having jurisdiction (AHJ) and, sometimes, they are just good practice. The following section discuss a few of them that are worth considering.

Geologic and Seismic Surveys

Naturally, geologic studies are required for pumped hydro and compressed air, but they are also good practices to conduct for other construction, especially when tall stacks of storage equipment need to be installed. A thorough understanding of the earth under the structures not only can help keep them intact but possibly can help to reduce costs by optimizing for the appropriate load and support. Geologic studies will determine more than just ground stability; they reveal the soil type and conductivity to help design a robust lightning protection system. Seismic studies are important not only to ensure structural integrity, but to help plan for the appropriate product compliance. Products that have significant weight and height may be susceptible to earthquake damage, and they should be tested and certified to withstand the expected conditions at the intended location. There are several seismic standards to consider: some are more appropriate for buildings, some for equipment, some for both, and others are relevant only to certain geographical zones.

Ecological Surveys

For pumped hydro, the ecological survey is necessary for procuring a permit to execute such a project. Additionally, ecological impacts are important for all construction projects, even smaller ones. Certain locations, for example, have sensitive wildlife concerns and must be factored into the initial site survey. Not paying attention to this could stop a project in its tracks midstream, causing major delays and cost overruns.

Public Use Studies

Public use studies are mostly relevant to pumped hydro projects, as many reservoirs have a double use for both recreation and storage. Understanding the nature of the public use will help determine how to operate the facility. Any amount of boating, swimming, and fishing in the waterways will limit the amount and rate that the storage can be used safely.

Regarding other energy storage types, some creative-minded individuals have thought of how their energy storage technologies can have a double use, not only to gain acceptance in the communities but to help pay for them as well. Using the energy storage as part of a resiliency program to ensure continuity of public safety services, provide backup power in residential centers, and collocate with data centers are some ways to get more use out of the energy storage. Some developers are even creating unique parkland and retail as part of their storage projects to assimilate them into the community where the storage is needed most. Batteries in an apartment building can provide backup power when the grid goes dark. Storage in an industrial park can reduce demand charges and provide resiliency. Energy storage near EV charging stations can limit the power quality problems associated with their huge power demands.

Thermal energy storage can have dual use as well. In northern countries, thermal energy storage is used extensively for winter heating. Some energy storage technologies can be designed to store and deliver both electrical and thermal energy when needed. Surveying the local communities' habits and latent needs will go a long way toward the public acceptance of a major energy storage project.

Environmental Impact Studies

Although energy storage does promote the extended use of renewable energy and is considered an enabler of a more environmentally beneficial power grid, it can hurt the environment if not executed well. Some of the factors to be considered are the carbon footprint of the energy storage technology, transportation of equipment to the site, and the construction materials, as well as emissions of the product while operating, and increased emissions of nearby power plants as a result of the storage's operation, light pollution, heat pollution, noise pollution, effect on waterways, effect of seismic activity from compressed air caverns, wildlife movement, and bird and bat navigation effects, among others.

No storage technology is completely green. Many batteries are made from materials that have significant effects on the Earth where they are mined and use extensive energy to manufacture. An overseas shipment of batteries could emit as much carbon as the battery itself weighs. Inefficient storage could cause local carbon emitters to emit even more carbon into the atmosphere to compensate for its round-trip inefficiency. CAES storage sometimes burns additional gas to regain the energy it stored underground. Even the seemingly benign pumped hydro technology has an impact on the environment. For example, for every kilogram of concrete in its dams and structures, a kilogram of carbon dioxide was emitted into the atmosphere during its production. Large dams like Hoover, Grand Coolie, and Grande Dixence used millions of tons of concrete in their construction.

Safety, Arc Flash, and Fire Hazard Analysis

Public safety is paramount to the success of any storage project. Batteries have been taking the brunt of recent attention in this regard. If the public deems a storage project unsafe, no amount of technical mitigation will win their acceptance of a nearby installation. Recent safety incidents have been caused by an assortment of errors from bad cell designs, manufacturing errors, system design shortcuts and oversights, sloppy installation practices, and just plain bad luck. As the industry learns from these mistakes, it will get better and, as a result, standards will be put in place to keep others from making the same mistakes.

Other technologies have their safety issues as well. For example, people have died in the construction and operation of pumped storage facilities. Just in 2000, the Grande Dixence in Switzerland suffered a burst penstock which triggered a mudslide that buried several buildings and people.

Third-party safety evaluations done at the product and site level are not only necessary in most jurisdictions but are also good for the project's success and the industry's reputation.

Grid Benefits and Transmission Studies

Ultimately, energy storage makes electricity more reliable. Since the grid is the main vehicle delivering electricity, utilities often require interconnection studies to ensure the installed energy storage not only improves the reliability but does not detract from it either. Utilities have developed models of their grid that they use to determine how the energy storage will interact with it under normal conditions and abnormal contingencies. If an energy storage system is controlled right, it will provide a positive benefit to the grid. Often, the controls are tuned to work with specific grids and need a few iterations resulting from this survey. In other cases, such a survey will reveal that transmission line upgrades are necessary to install energy storage at the desired location.

Meteorological and Environmental Effects Studies

Pumped hydro certainly requires a meteorological study to determine how the expected yearly rainfall patterns will affect its water supply and movement. Meteorological studies can benefit other energy storage projects as well. For example, installation progress can be upset by nonoptimal weather, enclosure finishes can be negatively affected by acid rain near an active volcano, and HVAC system performance can be affected by average local temperatures.

Archeological Survey

When the Three Gorges Dam was built in China, thousands of ancient shrines were in danger of being covered by water after it was filled. Although many were saved, some were lost. Around the globe, many other construction projects are halted when something is discovered during excavation. This is especially likely in historic urban settings but could happen anywhere. Knowledge of potential risks can help set expectations and defer project penalties.

Permits

Countless permits exist in local jurisdictions necessary for public acceptance of any project, from putting siding on a house to building a thousand-acre reservoir. Local AHJs found at the state and municipal government centers are good resources toward a successful project. Permits include the usual interconnection, electrical, building, construction, and land use, but can also include water use, Army Corps., waterways, transportation, conservation, and roadway permits, depending on the project type.

7.3.1.3 Installation

Of all the energy storage projects, the longest to install is pumped hydro storage. It is not unusual for them to take 10 to 15 years to complete. Sometimes, in order to accommodate thousands of workers and their families, whole villages, including churches, schools, stores, and recreation centers, are built near the project. Such was the case with the Hoover Dam in the United States (Figure 7.3) and the Grande Dixence in Switzerland (Figure 7.4).

Figure 7.3 Boulder City, NV in 1932. (Source: U.S. National Archives.)

Figure 7.4 Grande Dixence's 285m tall dam. (Source: ETH-Bibliothek Zürich, Bildarchiv/Fotograf: Comet Photo AG (Zürich)/Com_FC35-0006-177/CC BY-SA 4.0.)

It is also not unusual for large energy storage projects to cost billions of dollars to execute. Some installation work is very specialized, such as the hollowing out of large salt caverns hundreds of meters below the Earth's surface for CAES plants. Other installations are very straightforward, such as the erection of buildings and the placement of electrical equipment in the case of batteries (see Figure 7.5).

However, no matter what technology it is, the installation costs are not proportional to the cost of the energy storage. Instead, it is tied to the size of the physical plant and the number of activities associated with that plant's assembly.

Consider a battery comparison for a moment. Storage A is highly energy-dense so that 10 MWh can fit into one 53-ft container, while only 2 MWh of a less dense Storage B can fit into a 53-ft container. Both technologies are being considered for a 10-MWh project. Installation activities include site preparation, concrete foundations, wire trenches, wiring, battery installation, PCS installation, transformers, switchgear, and overhead. Storage B, in five containers, will require five times the physical effort than with Storage A's one container because there are five times as many containers, dirt to move, concrete to pour, trenches to lay, wires

Figure 7.5 Example of cookie-cutter simple foundation work on standard battery enclosures. (Source: Author.)

to pull, and connections to make. Not to mention the additional shipping, handling, lifting, placing, inspecting, testing, assembly, and the myriad of other activities associated with their installation.

Installation is not just a simple ratio of the energy storage cost. More often, it is the opposite. Furthermore, installation costs are probably the most unpredictable part of any project cost plan. There are so many factors involved that even the most seasoned industry veteran can be caught off guard. That is why many manufacturers of storage technologies are going in the direction of making their products modular (Figure 7.6). The less site work, the less cost impact its variability will have on the project.

7.3.1.4 Summary of Up-Front Costs

Storage equipment, surveys, inspections, permits, engineering, site-work, foundations, transportation, power equipment, rigging, assembly, wiring, plumbing, testing, and commissioning are all part of the tangible up-front costs. For large bulk energy storage, such as pumped hydro and CAES, thermal, liquid air, and some forms of flow batteries, site work is the dominant portion of the installation costs. Energy storage using batteries and flywheels tends to have higher equipment costs but less on-site construction costs. Newer generations of battery storage products are

Figure 7.6 Example of premade factory-built modules that get dropped in place and wired. (From: [2]. Reprinted with permission.)

becoming more modular, which allows the modules to be dropped in place and commissioned with little ground effort (Figure 7.6).

One of the intangible costs is that associated with planning for future capacity loss over a project's service life. Many customer contracts require an expected energy capacity over the life of the product. Since many battery chemistries will degrade over time and use, plans for how to make up lost capacity throughout the service life must be made up-front, and not as an afterthought years down the road. Preparing for expected capacity loss include the extra land, larger concrete pads, and additional wiring and infrastructure adds to the up-front construction and installation costs.

7.3.2 Operating Costs

$$LCUS = \frac{Capex_0 + \boxed{\sum_{t=1}^{T} \frac{Opex_t}{(1+r)^t}}}{\sum_{t=1}^{T} \frac{Energy\_discharged_t}{(1+r)^t}}$$

In most energy storage projects, the up-front cost is by far the largest percentage of the total cost of ownership, so it is not surprising that the ongoing maintenance and operating costs are deemphasized when comparing different technologies. Nevertheless, round-trip losses and self-discharge will inevitably reduce operating profits. Furthermore, as the prices of energy storage, particularly of battery storage, come down, it is possible, and even probable, that the service costs associated with round-trip energy losses can be as large or larger than the up-front cost of the equipment.

7.3.2.1 Charging Costs

Like income taxes, round-trip energy losses are subtracted from the gross amount of energy put into storage, becoming lost and unusable. Unfortunately, the storage operators have to pay for the gross charging energy. For example, if over 20 years an 80% round-trip efficient storage system is cycled every day and is assessed \$0.05 per kWh for its charging power, its charging costs will be:

$$\begin{aligned} Cost_{Charging} &= \#Cycles \cdot Losses_{RT} \cdot Cost_{kWh} \\ &= 7300 \cdot 20\% \cdot \$0.05 = \$73/kWh_{del} \end{aligned} \tag{7.4}$$

A counterargument to this cost is that in some power purchase agreements (PPA), particularly those associated with PV energy plants, the charging from the battery is solely from the PV array, not the grid, so its charging energy is purportedly free. However, looking at the larger picture, assuming the entire planet is powered from PV arrays during the day and batteries at night, if the batteries lose 50% of the energy put into them each day, it would be necessary to size the PV arrays (and the batteries) 150% greater than if the batteries had been 100% efficient. Therefore, PV charging energy is not free. Somewhere down the line, someone has to pay extra for a larger PV array to account for the additional losses in the storage facility.

7.3.2.2 Auxiliary Power

Speaking of taxes, there are also electric bills that need to be paid. The auxiliary ac power lights the enclosures, runs the controls, sounds the alarms, powers the HVAC and heaters, pumps electrolyte and coolant, and drives air blowers. It is not the most significant of costs, but it can add up. As with residential and commercial electric bills, air-conditioning is the biggest consumer. If the battery is enclosed in an insulated box, its self-generated heat needs to be removed, or its temperature will rachet up until it fails. For example, if a battery is 90% round-trip efficient, 10% of its energy will be lost to heat during each cycle. To remove that heat, an air-conditioning system will require about one-third of the heat energy in electricity (COP = 3). In this example, the auxiliary power will be 3.33% of the discharged energy to the grid. If the auxiliary power costs $0.15 per kWh (retail rate), the cost per cycle will be $0.005 per kWh, or $1.8 per discharged kWh in a year and $36/kWh for a 20-year service life. There are many variables in this example, such as the cost of electricity, the storage round-trip efficiency, and the number of cycles per year. A simple formula to estimate what the cost of auxiliary power for cooling the batteries will be is:

$$Cost_{Aux\_Cooling} = Cost_{kWh} \cdot \#\_cycles \cdot \frac{(1-\eta_{RT})}{COP_{HVAC}} \tag{7.5}$$

Where COP_{HVAC} is the coefficient of performance of the HVAC system, η_{RT} is the round-trip efficiency of the batteries and wiring inside the enclosure that needs to be cooled, and $Cost_{KWh}$ is the cost per kWh paid for auxiliary power. If an energy storage technology does not require 25°C and can operate mostly at temperatures that are above the outside ambient temperature, then the cooling costs will be reduced by using a more efficient air-to-air heat exchanger, or simply an air exchanger. High-temperature batteries are one example where active HVAC is not required.

Another large auxiliary power drain is specific to flow batteries. Flow batteries work when their electrolytes are flowing, thus they require pumps to get them started and keep their power going. This can be as much as 10% of the total power of the flow battery. Because the stack does not output any power until there is liquid flow, the pumps are powered by the auxiliary power circuit. A 5% power drain on the stack when it is operating will certainly eat into the round-trip efficiency numbers but is exacerbated if the pumps continue to run when the stack is not cycling. However, if the battery is operated in a scheduled cycle mode, the pumps can be turned off between cycles, allowing the electrolyte to drain out of the stacks and back into the tank. This incurs a delayed start-up when the cycle begins, but since the cycles would be scheduled, the pumps can be started a few minutes before the expected power delivery.

7.3.2.3 Maintenance Costs

Another cost that cannot be ignored is related to maintenance. No energy storage technology is immune to nature's tendency toward increased entropy. Water turbines cavitate or wear out, compressed air pipes corrode, bearings fail, cables fray, pumps seize, pipes crack, chemicals degrade, and batteries lose capacity. All of the longest-lasting energy storage technologies require some continual maintenance to keep them running over many years. Even pumped storage water reservoirs eventually fill with silt and must be dredged periodically to maintain their capacity. Flow batteries' electrolytes purportedly last forever, but they also have to be refurbished on a regular cadence. Like an old tractor, pumps, motors, filters, and stacks require regular maintenance to keep them running through their warranty period.

Any system with power electronics will also have to be serviced at some point. Electronic circuits are known to have a finite service life depending on the design and quality of their components. Thermal cycles, air contaminants, corrosion, vibration, metal migration, and cosmic rays all slowly degrade the connections between components in even the most robustly designed circuit. Eventually, statistics wins, and they need to be replaced.

Air-conditioning systems are also a major service item. Most manufacturers will not deliver more than a 7-year commercial warranty for a packaged HVAC system. In addition, HVAC systems require yearly maintenance to replace filters and inspect components.

Storage projects that require a minimum amount of energy storage over their entire service life and use technology that degrades over time, may require a continual update of its capacity if not oversized enough at the start. For example, if the degradation rate is 40% over 20 years, then 40% of a site's base capacity will have to be added in 20 years. Some battery vendors include a capacity-maintenance plan in their warranty costs to cover this.

Energy storage requires monitoring and management. Some plants like PHS and CAES may require one or more operations personnel on site because there are so many moving parts (the Grande Dixence has 24); however, others, such as a battery-based system, may not need anyone on-site. In fact, most battery energy storage sites are monitored remotely until an alarm lights up. Some manufacturers are implementing artificially intelligent monitoring to allow more autonomy, as learning algorithms watch for trends in certain data streams to predict failures and maintenance cycles.

7.3.2.4 Other Ongoing Costs

All equipment manufacturers offer a short one to 3-year warranty that comes with the price of their equipment. For longer guarantees of desired performance, owners can purchase long-term warranties that will protect their investment from any unexpected failures or departures from specification. These plans typically are 1% to 5% of the original purchase price per year, depending on the inherent reliability and/or notoriety of the components. For example, a power transformer will probably have less than 1% per year added cost, while an air conditioning unit will have 5% or more. Power conversion systems are somewhere in between, depending on the manufacturer and risk tolerance.

Batteries have two or more kinds of warranty. At the basic level, they have a break/fix warranty that covers the battery if it utterly fails. In addition, as already mentioned, manufacturers may offer capacity guarantees which warrant that the capacity degradation will be no more than a given projection for a specified duty profile. Furthermore, there are warranties for availability, efficiency, and other metrics.

7.3.3 Summary of Costs

The costs buckets described in this chapter, both up-front and ongoing, tangible and intangible, all detract value from the energy storage asset. To give some perspective of the relative costs of energy storage, the following are a couple of representative examples of PHS and battery-based energy storage.

Pumped hydro facilities are enormous projects and can have billion-dollar price tags. Raccoon Mountain's 36 GWh cost $310 million in 1978 [3], $1.3B in 2021 dollars, making it $33/kWh. The 24-GWh Bath County PHS in VA cost $1.7B in 1985 [4] when it was complete, which in today's dollars is about $4.2B, making its up-front cost $177/

kWh. Batteries are currently more expensive: the hardware alone is in the \$150 to \$250 per kWh range as of 2021. Project installation and associated balance-of-plant push the total installed cost from \$250 to \$350 per installed kWh. Nevertheless, 10 years ago, the batteries were easily 10 times these values, so one can expect steady improvements ahead.

7.3.4 Delivered Energy Basis

$$LCUS = \frac{Capex_0 + \sum_{t=1}^{T} \frac{Opex_t}{(1+r)^t}}{\boxed{\sum_{t=1}^{T} \frac{Energy\_discharged_t}{(1+r)^t}}}$$

The cost of energy storage with respect to its delivered energy is a meaningful way to compare energy storage technologies and projects that cycle regularly. The implied value is the energy discharged regularly from storage. A cautionary note: it still can be skewed to favor one technology over another. For example, if a particular energy storage technology has little cycle life degradation but has a short calendar life, the conditions can be set to cycle continuously and unrealistically throughout the day and evaluated over a short service life. This will maximize the denominator, minimize the numerator, and result in a very low LCUS compared to other technologies running the same profile. To be fair, the assumptions and conditions have to be close to that of a real-world application, and explicitly stated as such.

7.3.5 Simple LCUS Exercises

To get a sense of the LCUS for various storage projects, a simplified analysis can be performed using estimates and assuming constant values for the terms in the equation. Specifically, if the operating and maintenance (O&M) cost per kWh, and delivered energy are constant throughout the service life, the LCUS can be simplified to:

$$LCUS = \frac{Capex_0 + \sum_{t=1}^{T} \frac{Opex_t}{(1+r)^t}}{\sum_{t=1}^{T} \frac{Energy\_discharged_t}{(1+r)^t}} = \tag{7.6}$$

$$LCUS = \frac{Capex_0}{kWh/yr \cdot \sum_1^T \frac{1}{(1+r)^t}} + Cost_{kWh} \cdot (1 - \eta_{RT}) + \frac{Cost_{O\&M}}{kWh/yr}$$

The first term, $\frac{Capex_0}{kWh/yr \cdot \sum_1^T \frac{1}{(1+r)^t}}$, converges to a fixed number at T = infinity.

The second term is constant, assuming the cost of electricity is constant.

The third term is also constant, assuming the operating and maintenance cost and the delivered energy per period are constants.

7.3.5.1 Pumped Hydro Storage

Consider Raccoon Mountain PHS, which has been running for the past 43 years. Below is a simplified LCUS using representative costs to give some perspective:

- Up-front (2021 value): $1.3B (was $310M in 1978);
- Charging cost: Assume $0.05 (bulk rate)/charging kWh;
- Efficiency: Assume 75% round-trip;
- Delivered Energy: Assume 4 hours per day @ 1.6 GW = 6,400 MWh = 6.4E6 kWh;
- Operating and Maintenance cost: Assume $10M/year;
- Discount rate: 7%.

Simplified example LCUS using Raccoon Mountain's up-front, charging costs, and O&M costs:

$$LCUS_{RM} = \frac{Capex_0}{kWh/yr \cdot \sum_1^T \frac{1}{(1+r)^t}} + Cost_{kWh} \cdot (1-\eta_{RT}) + \frac{Cost_{O\&M}}{kWh/yr}$$

$$LCUS_{RM} = \frac{\$1.3B}{6.4E6 \cdot 365 \cdot \sum_1^T \frac{1}{(1+0.07)^t}} + \$0.05 \cdot (1-0.75) + \frac{\$10M}{6.4E6 \cdot 365}$$

$$LCUS_{RM} = 0.053 + 0.0125 + 0.0043 = 0.069 (T = 20 \text{ years})$$

$$LCUS_{RM} = 0.041 + 0.0125 + 0.0043 = 0.058 (T = 43 \text{ years})$$

The plot of this LCUS with respect to the length of service is shown in Figure 7.7. Notice that the LCUS converges to a constant of \$0.06 after year 20.

7.3.5.2 Lithium-Ion Batteries

Batteries have a higher LCUS because their up-front costs are higher, and they do not last as long. For the following example, assume that the battery is oversized by 150% to make it last 20 years. Therefore, the battery's daily cycle depth will be 67%. Also assume 93% RT efficiency at the ac POI. The O&M will be about 3% of the up-front equipment costs,

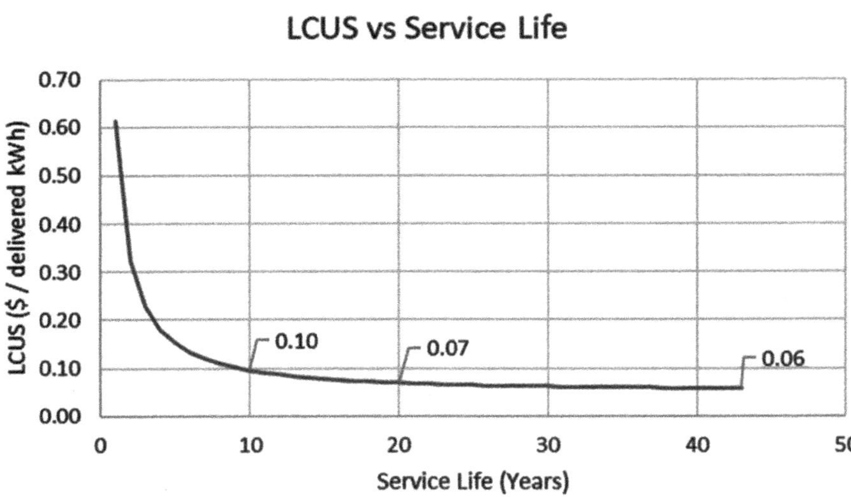

Figure 7.7 LCUS vs service life.

dominated by the performance warranty costs of the battery and balance of plant equipment.

- Up-front (2021 value): \$350/installed kWh;
- Charging cost: assume \$0.05 (bulk rate)/charging kWh;
- Efficiency: assume 93% RT;
- Delivered energy: assume 67% × installed kWh each day;
- Maintenance cost: Assume \$5/installed kWh-year;
- Discount rate: 7%.

Simplified LCUS using typical lithium-ion up-front and charging costs:

$$LCUS_{LiB} = \frac{Capex_0 / kWh_{installed}}{DoD_0 \cdot \sum_1^T \frac{1}{(1+r)^t}} + Cost_{kWh} \cdot (1 - \eta_{RT}) + \frac{Cost_{O\&M} / kWh_{installed}}{DoD_0 \cdot 365}$$

$$LCUS_{LiB} = \frac{\$350/kWh_{installed}}{0.67 \cdot 365 \cdot \sum_1^{20} \frac{1}{(1+0.07)^t}} + \$0.05 \cdot (1 - 0.93) + \frac{\$5/kWh_{installed}}{0.67 \cdot 365}$$

$$LCUS_{LiB} = 0.135 + 0.0035 + 0.02 = 0.16 (T = 20 \text{ years})$$

The first term, \$0.135, is much larger than the PHS \$0.041 relative up-front cost because of the \$350 per installed kWh price tag and the shorter service life. If the batteries do not have to be oversized, they last 40 years, and they cost \$200 per installed kWh, their LCUS will be close to that of PHS. Notice also that the charging cost for the PHS is almost a third as much as its up-front cost on a per-delivered kWh basis, whereas the battery's charging cost is a much smaller fraction. Another way to improve the LCUS of the battery would be to augment the system as it degrades over time and not oversize it so much in the beginning. This would shift some of the up-front costs to the O&M term and reduce the LCUS by about 10%.

7.3.5.3 Flow Batteries

Some flow battery technologies have the advantage that they do not have to be oversized to account for capacity degradation. Consider an example flow battery with infinite cycle life and a rough up-front cost of \$250 per kWh. It has a lower energy density than lithium-ion, so its installation costs will be higher, leading to a total up-front cost of $450 per kWh. Its round-trip efficiency is 65%, and yearly service costs are 5% per year of its equipment costs because of insurance, parts replacements, and warranty. It does, however, deliver all of its energy daily.

- Up-front (2021 value): $450/installed kWh;
- Charging cost: assume $0.05/charging kWh;
- Efficiency: assume 65% RT;
- Delivered energy: assume 100% x installed kWh each day;
- Maintenance cost: assume 5% of $250/installed kWh-year;
- Discount rate: 7%.

$$LCUS_{Flow} = \frac{Capex_0/kWh_{installed}}{DoD_0 \cdot \sum_1^T \frac{1}{(1+r)^t}} + Cost_{kWh} \cdot (1-\eta_{RT}) + \frac{Cost_{O\&M}/kWh_{installed}}{DoD_0 \cdot 365}$$

$$LCUS_{Flow} = \frac{\$450/kWh_{installed}}{1 \cdot 365 \cdot \sum_1^{20} \frac{1}{(1+0.07)^t}} + \$0.05 \cdot (1-0.65) + \frac{\$12.5/kWh_{installed}}{1 \cdot 365}$$

$$LCUS_{Flow} = 0.116 + 0.0175 + 0.034 = 0.17\left(T = 20 \text{ years}\right)$$

This is slightly higher than lithium-ion because of higher charging service costs. The service life was set to 20 years to reflect the typical manufacturer's specifications. A longer life is possible since the electrolyte does not wear out, but stacks, pumps, pipe fittings, and valves do. Perhaps the products could be manufactured in such a way that wholesale replacement of the major components will sustain the product over a much longer term than 20 years. If that is the case, the LCUS will decrease and compete with lithium-ion.

7.3.6 Applications of Energy Storage that Make Cents

Regardless of how one technology stands up against another, ultimately, the energy storage needs to return a profit in the application it is used. If the project is chasing a $0.05 per kWh price spread in a daily arbitrage market, then its total costs per cycle has to be less than \$0.05 per delivered kWh, including up-front, ongoing, and charging costs. Lithium-ion batteries are not there yet, but they are getting close. There are other applications in which energy storage does make a profit. For example, the Hornsdale application in Figure 6.33 has wider price variability and can afford a higher battery cost. Chile is unique in that power plants are required to reserve the top 5% of their power range to be able to respond to frequency deviations in order to stabilize the grid (Figure 7.8).

If the bulk energy price is \$50/MWh, and a power plant size is 400 MW, its lost revenue each year would be:

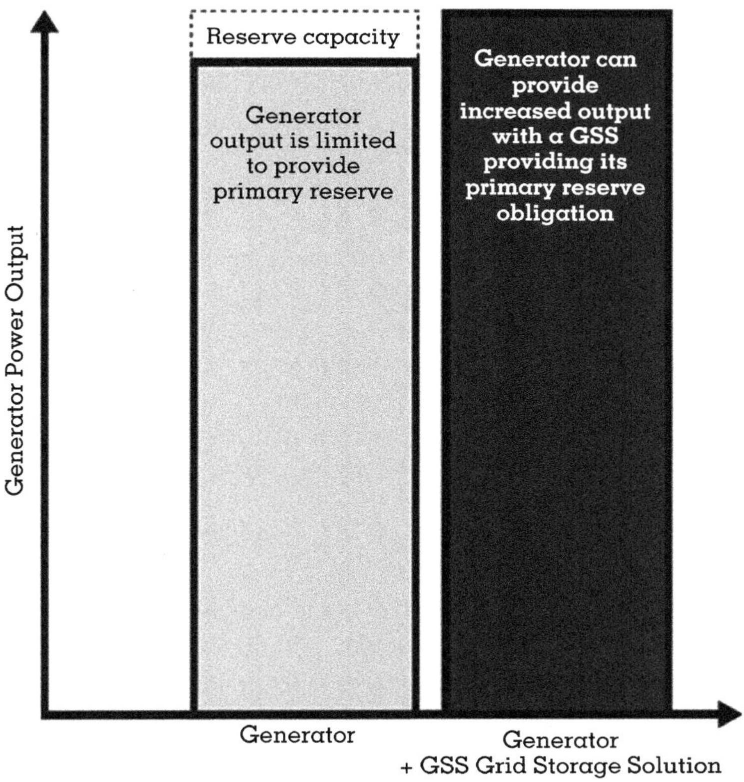

Figure 7.8 In certain power markets, power producers are required to reserve 5% of their operating capacity to respond to primary frequency reserve commands. (Source: NEC. Reprinted with permission.)

$$R_{Lost} = 400MW \cdot 5\% \cdot 8760hr/yr \cdot \$50/MWh = \$8.8M/yr \quad (7.7)$$

To allow the power plant to run at full capacity, a battery can be used to provide the primary frequency response. For 5% of 400 MW, a 10-MW battery with 20 minutes of run time (10 up and 10 down) can be deployed to respond to primary frequency events. In this unique application, the increased revenue of $8.8M far outweighs the costs of the battery. More than just a financial benefit, the faster reacting battery systems provide better stability to the grid than the slower and less precise response from a large steam turbine power plant.

Many other markets exist around the world in which energy storage technologies are economical and even profitable for their owners. There will always be opportunities for energy storage to solve temporary problems, such as deferring investment in a crowded transmission corridor, deferring installation of an undersea cable to a vacation island, reducing price differentials in undersupported or imbalanced grids, and others. As these markets get served by enough energy storage to stabilize the grid and the markets, the price differentials fall and discourage additional newcomers. Ultimately, the converging cost target is the market electricity price minus the costs of generation, transmission, distribution, and other costs. At the wholesale level, where gigawatt power providers deliver power generation and storage, price targets are in the few cents range. Energy storage at the local level, subtracting out transmission and distribution costs, could have more leeway in its targets. As energy storage becomes more and more pervasive and less expensive, this may naturally happen by itself.

7.3.7 Sum of the Parts

Improving energy storage effectiveness is not just about improving the technology itself. Significant benefits can be realized by optimizing where the energy storage is placed on the grid and how it interacts with the loads and generation.

7.3.7.1 Storage Location

Most storage is placed at convenient locations in the grid that are capable of handling its power without negatively affecting power quality, such as in an existing switchyard or substation. Selecting a location near the cause of variable power will result in the lowest losses and controlled grid conditions. Some illustrations can help show how. Figure 7.9 shows an urban load powered by a distant power plant. The urban load has daily power fluctuations, peaking in the day and trenching at night.

Since wire losses are a function of the square of its load current, the power losses during two-times peak-power periods can be four times the daily average. Placing energy storage near the load center reduces the peak currents on the long transmission line and keeps the peak power stresses on a much smaller part of the grid (Figure 7.10). This was successfully done in the Ni-Cd GVEA BESS and the lithium-ion Atacama Desert installations discussed in Chapter 6.

Additionally, when renewable power is introduced into the grid, it also brings cyclical and somewhat random power profiles to the transmission network, which causes excessive power losses due to its peaks (Figure 7.11).

Putting energy storage near the disturbing forces reduces the peak currents on the grid, lowers the losses, and improves the power quality (Figure 7.12). This was successfully done in the lithium-ion Maui and

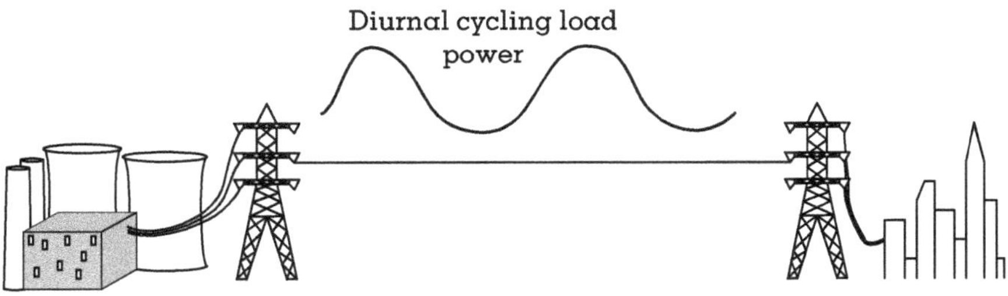

Figure 7.9 Diurnal load profiles can increase losses on long transmission lines.

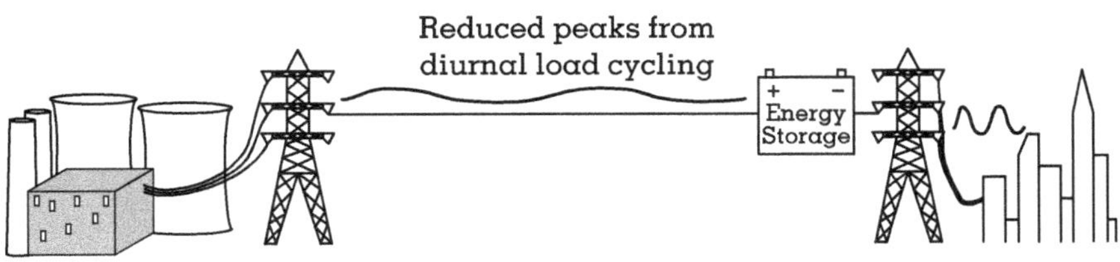

Figure 7.10 Energy storage positioned near the load centers can decrease losses on long transmission lines due to diurnal load profiles.

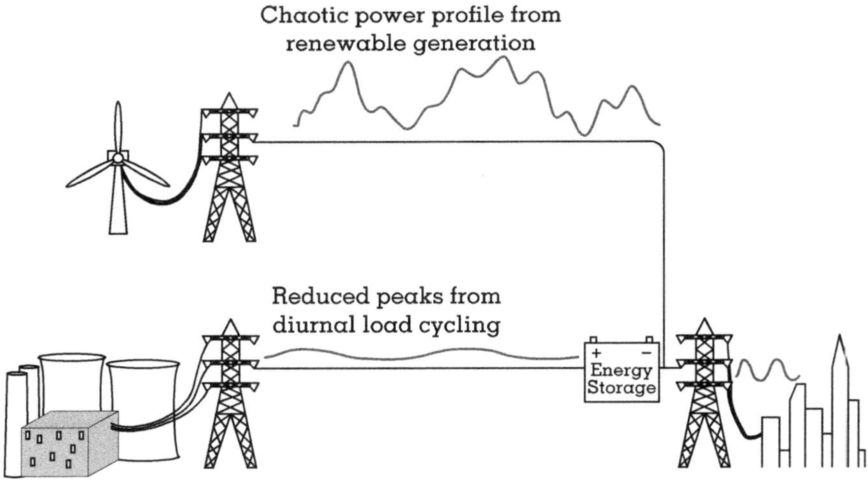

Figure 7.11 Renewable generation profiles can increase losses on long transmission lines due to high peak power to average profiles.

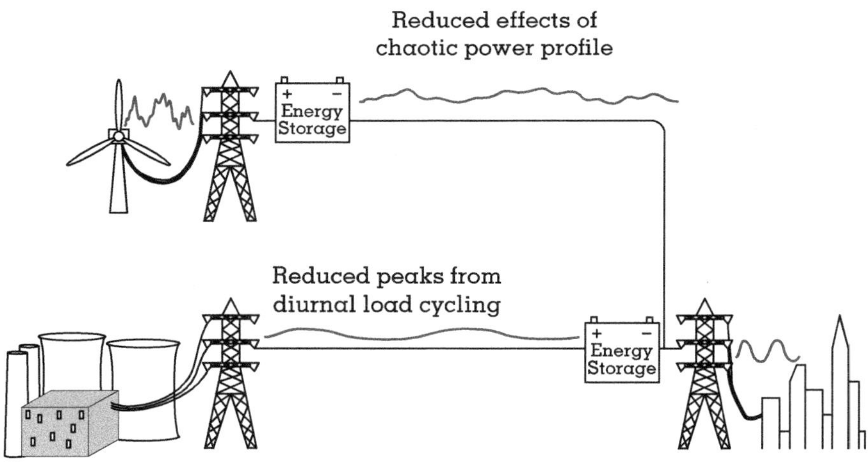

Figure 7.12 Energy storage placed near renewable generation can decrease transmission line losses.

Laurel Mountain projects discussed in Chapter 6, and also in many installations where storage is sited right next to large PV plants.

Unfortunately, nothing is ever this simple. Grids are complex, and picking the right location is always a challenge, especially with large bulk power technologies like pumped hydro and CAES. Smaller modular energy storage can be distributed throughout the grid to help reduce the losses in more places. Modeling the grid, its generation, and load profiles, along with the addition of energy storage with an eye on the resulting power flow efficiency, can help to determine the best place to put the resources.

7.3.7.2 Storage Interaction with Generation and Loads

The classic standard of energy storage connected to a single point on the grid where it absorbs and delivers energy as needed is certainly the most straightforward, but it may not be the most efficient implementation. Where this standard is applied, a significant amount of energy is diverted into and out of storage every day going through several steps of energy conversion each time (Figure 7.13). This extra loop taken by the peak energy adds additional losses to the system.

A tighter integration between energy storage, loads, and generation can improve the overall efficiency of the storage process. In fact, this is already done where long-term energy storage is integrated into

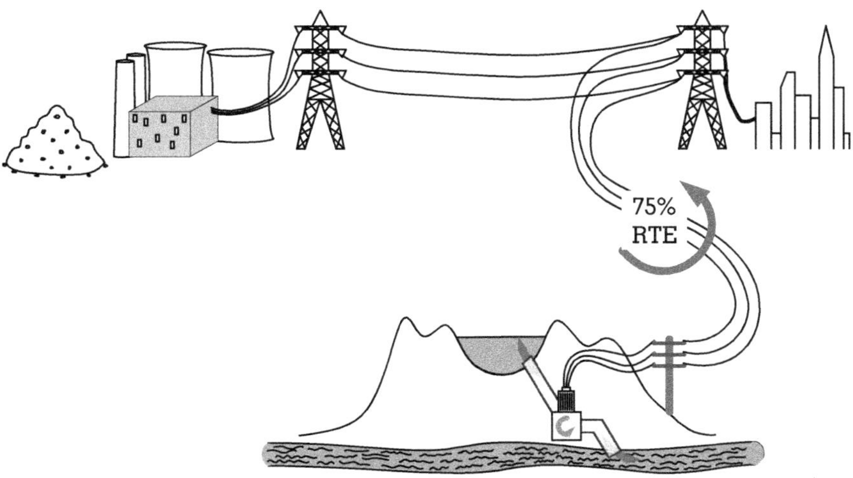

Figure 7.13 Classic energy storage incurs an additional round-trip energy loss.

hydroelectric generation facilities using a dammed reservoir. Battery energy storage is increasingly being installed on the dc side of a PV site's dc/ac power conversion to increase the overall efficiency between the PV to ESS charging process. In another example, thermal energy is stored in concentrated solar plants, so that the plants can provide continuous power to the grid despite being warmed only during the sunny parts of the day (Figure 5.18). A similar process can be implemented in a nuclear or fuel-burning plant to allow a constant thermal process to support time-varying loads without too much stress (Figure 5.21). These examples of energy storage do not incur any additional round-trip losses of the classic storage model. In yet another example, the large number of synchronized spinning machines in CAES plants can be used for local inertial response providing another valuable source of energy storage.

On the load side of the equation, tight integration of renewable energy generation with non-time-critical loads such as desalination, electrolysis, refrigeration, and heat can reduce the grid's required energy storage. For example, desalination processes can be powered whenever there is sun or wind, while H_2 production can occur whenever power is available as well. Thermal energy storage for domestic heating and cooling can be coordinated with renewable energy production. All of these loads are not time-critical because each entails its own storage mechanism, whether thermal, liquid, or gas; so, they can be powered intermittently without too much service disruption. This coordination between load and supply can be either directly coupled (e.g., a solar plant directly powering a desalination plant), or it can be virtually coupled by network protocols. In the latter case, intermittent generators connected to some remote corner of the grid, can communicate to local, non-time-critical loads their output power which can then be used to control the load's consumption. Such coordination can increase the pool of system-wide, energy storage for use by other time-critical loads.

7.4 Concluding Remarks

Energy storage has come a long way from when it was accidentally discovered, to becoming an integral part of our daily lives, and to the role it will play to help humans thrive in harmony with our planet. This book has described the main energy storage technologies that have the biggest near-term impact. However, future energy storage technologies using novel chemical and new electrochemical storage means will certainly arise and challenge the incumbents.

This book has presented objective ways to look at any storage technology, existing and new, and showed how their individual merits can be compared on an equal basis of true value. LCOS is useful for evaluating energy storage for backup and capacity applications, while LCUS is useful for evaluating energy storage for repetitive cycling applications. In both metrics, size, efficiency, cycle-life, calendar-life, confidence, reliability, and solid designs all affect the total cost of ownership.

Using these tools shows how some energy storage products are more suited for some applications, others less so, and that there is no one energy storage that fits every application. Thermal energy storage is a good fit when tightly coupled to a thermal generation plant. Pumped-hydro is ideal when land, height, and water are plentiful. Mechanical energy storage works great for high cycle life applications and where space is available, and batteries with their wide range of capabilities fit a variety of applications, big and small.

Despite its costs, when applied appropriately, energy storage can increase the effectiveness of renewable energy resources, replace fuel-burning peaker plants, increase the efficiency of existing infrastructure, and even improve the reliability of power delivery to customers.

The future of energy storage is certainly going to include new generations of electrochemical, chemical, mechanical, and thermal storage technologies, but it will also bring more advanced controls, and tighter integration with generation and loads; all of which will have a huge potential to amplify its effectiveness. Stay tuned.

References

[1] Sears J. R., "Thermal and Compressed Air Storage (TACAS) Technology: The Road to Commercialization," Active Power.

[2] Boll R., "Less Than a Year after Introduction, Fluence's Sixth-Generation Technology Is in the Field," April 26, 2021, https://blog.fluenceenergy.com/fluence-sixth-generation-energy-storage-technology-deployed.

[3] TVA, "The Mountaintop Marvel," https://www.tva.com/about-tva/our-history/built-for-the-people/the-mountaintop-marvel.

[4] Lewis S., April 11, 2018, "The 10 Largest Pumped-Storage Hydropower Plants in the World," *Engineering News-Record,* https://www.enr.com/articles/44302-the-10-largest-pumped-storage-hydropower-plants-in-the-world.

List of Acronyms, Constants, and Terms

18650 A 65-mm long cylindrical cell with an 18-mm diameter

21700 A 70-mm long cylindrical cell with a 21-mm diameter

26650 A 65-mm long cylindrical cell with a 26-mm diameter

4680 A 80-mm long cylindrical cell with a 46-mm diameter

50100 A 100-mm long cylindrical cell with a 50-mm diameter

60138 A 138-mm long cylindrical cell with a 60-mm diameter

A Ampere: a unit of current equal to one coulomb per second

A123 A123 Systems: a manufacturer of lithium-ion batteries and systems

ABB International large equipment supplier based in Switzerland

ac Alternating current: current that alternatively flows in both directions in a conductor

AES Applied Energy Services: an IPP and early promoter of lithium-ion batteries in grid applications

AGC Automatic generation control: a signal sent from a central controlling entity to set the power output of multiple generators

AGM Absorbed glass mat: a fiberglass pad saturated with electrolytes sandwiched between two plates in a lead-acid battery

ANR American Natural Resources: a company that formed a JV (joint venture) with ECD to create Ovonics, a company that commercialized the NiMH battery

bar Unit of pressure equal to 14.5 PSI and 100,000 Pa

BESS Battery energy storage system

BOL Beginning of life—with respect to the characteristics of an energy storage medium at the point of its installation.

BYD Build Your Dreams: a Chinese manufacturer of batteries, energy storage systems, and electric vehicles

CAES Compressed air energy storage: energy storage using the pressurization of gas as a means of stored potential energy

CAISO California ISO (*see* ISO)

Capex Capital expense: the up-front costs of installing an operating, asset including equipment and installation costs

CC Constant current: a method of charging or discharging a battery wherein the battery's current is held constant

cm Centimeter: a unit of length equal to one one-hundredth of a meter = 0.39 inches

COP Coefficient of performance: a metric describing the ratio of heat moved vs electricity input into an HVAC system

c_p Specific heat capacity

CP Constant power: maintaining a constant power during discharge or charge

CP/2, CP/4, ... Discharge and charge rates defined by a ratio of the nominal energy capacity (e.g., CP/2 is a rate such that the cell is delivering or accepting a power equal to the nominal energy of the cell divided by 2). In an ideal world, this would result in a discharge time of 2 hours, but because of losses, it will be less.

CSFM Converter-fed synchronous machine: a coupling of a full-power double conversion ac/ac power converter with a standard synchronous machine to achieve full frequency and power control

CSP Concentrated solar power plant: concentrates large areas of solar power on a small focal point to raise a material's temperature

CV Constant voltage: maintaining a top-of-charge condition by holding the terminal voltage constant

dc Direct current: current that flows in one direction in a conductor

DFIM Doubly-fed induction machine: an electric machine capable of motoring or generating at a different frequency than the ac grid by injecting variable power into the rotor windings

DoD Depth of discharge: percentage of charge discharged with regard to the total dischargeable capacity

DOE United States Department of Energy

ECD Energy Conversion Devices: a company that developed NiMH technology

ELDC Electrolytic double-layer capacitor

EMS Energy management system: a control system that manages the components of an ESS following predefined goals of performance, availability, service-life, and others

EOL End of life: with respect to the conditions of the energy storage medium, when it is no longer fit for service, or other predefined termination criteria

EOS The name of a company manufacturing and selling zinc-carbon batteries

EPRI Electric Power Research Institute

ERCOT Electric Reliability Council of Texas (see ISO)

ESD Electrostatic discharge: a high voltage arc or discharge of an accumulation of charge

ESS Energy storage system

ESS Inc. The name of a company manufacturing and selling all-iron flow batteries

EV Electric vehicle: any type of vehicle that is powered by energy stored in a battery

F Faraday constant = 26.8 Ah / mol

F Farad: unit of capacitance

FR Frequency regulation or response

GE General Electric: large U.S.-based company that specializes in turbines, engines, electric machines, and other power equipment

GESDB DOE's global energy storage database

GPS Global positioning system: a satellite-based geo-positioning system that has a side benefit of transmitting an extremely accurate global time reference

GVEA Golden Valley Electric Authority: power cooperative near Anchorage, Alaska

GW Unit of power: equal to one billion watts

GWh Gigawatt-hour: unit of energy equal to one billion watt-hours

HVAC Heating and cooling system for regulating temperature of the ESS components

Hz Hertz: a unit of frequency in cycles per second

i^2R Power losses in a conductor due to imperfect electron conductance

IEA International Energy Association

IPP Independent power producer

ISO Independent system operator: collection of utilities regulating power system quality and non-discriminatory access to regional transmission. (In the United States, there are NYISO, CAISO, PJM, MISO, SPP, ISO-NE, and ERCOT.)

ISO-NE New England ISO (see ISO)

J Joule: unit of energy equal to one watt times one second

JV Joint venture

kg Kilogram: unit of mass equivalent to 2.2 pounds at sea-level

kJ Kilojoule: unit of energy equal to 1,000 joules

km Kilometer: a unit of length equal to 1,000 meters = 0.62 miles

kph Kilometers per hour: unit of speed = 0.62 mph

kW Unit of power: equal to 1,000W

kWh Kilowatt-hour: unit of energy equal to 1,000 Wh

l Liter: unit of volume equal to one-thousandth of $1 m^3 = 0.26$ gal

LAES Liquid-air energy storage: a method to store energy in the form of liquid gas

LCO A variant of lithium-ion chemistry using a cobalt oxide cathode

LCOE Levelized cost of energy: a metric to determine cost of providing energy (applicable to power plants)

LCOS Levelized cost of storage: a metric to determine the cost of having access to energy storage of a certain energy rating

LCUS Levelized cost of using storage: a metric to determine the cost of using energy storage with respect its delivered energy

LFP A variant of lithium-ion chemistry using an iron-phosphate cathode

LMO A variant of lithium-ion chemistry using a manganese-oxide cathode

LNO A variant of lithium-ion chemistry using a nickel-oxide cathode

LTO A variant of lithium-ion chemistry using a lithium-titanate anode

m Meter: a unit of length equal to 3.2 ft

MECO Maui Electric Company: a utility company in Hawaii

μF Microfarad: unit of capacitance equal to one millionth of a farad

MISO Midwest ISO (*see* ISO)

MIT Massachusetts Institute of Technology

MJ Megajoule: unit of energy equal to 1 million joules or 278 Wh

mm A unit of measure equal to one-millionth of a meter

mm Millimeter: unit of length equal to one-thousandth of a meter

mph Miles per hour: unit of speed

mTorr A measure of pressure usually for vacuum since 1 mTorr = 19 mPSI

MW Unit of power: equal to 1 million watts

MWh Megawatt-hour: unit of energy equal to 1 million Wh

MWHTP Megawatt-hour throughput: a measure of how much energy has passed in and out of a storage facility

Opex Operating expenses: the ongoing costs of operating an asset (such as fuel, maintenance, warranty, insurance, taxes, leases, interest, etc.)

N Newton: a unit of force

Nadir Lowest point

NCA A variant of lithium-ion chemistry using a nickel cobalt aluminum oxide cathode

NEC Nippon Electric Corporation: a Japanese manufacturer of many electronic computers, communications equipment, and batteries and energy storage systems

NEC ES NEC Energy Solutions: a part of NEC that built and sold energy storage systems

Ni-Cd Nickel-cadmium

NiMH Nickel-metal hydride

NMC A variant of lithium-ion chemistry using a nickel manganese cobalt oxide cathode

NREL National Renewable Energy Laboratories

NYISO New York ISO (see ISO)

OFLS Over frequency limit setting: a relay setting that will trip a breaker if the grid frequency goes above this value

Pa Pascal: a unit of pressure equal to $1 N / m^2$

PCS Power conversion System: an electronic system that converts dc power to ac, and, when so equipped, ac to dc

PFR Primary frequency response: a power generator or storage asset that can respond within 30 seconds of a grid's frequency deviation

PHS Pumped hydro storage: energy storage using the elevation gain of water as a means of stored potential energy (note: some literature uses PSH for pumped storage hydro instead)

PJM Pennsylvania, Jersey, Maryland ISO (see ISO)

PPA Power purchase agreement: an agreement between a utility and an IPP that defines how power and energy is to be delivered (and under what conditions) and the compensation for those services

PSI Pounds per square inch: a unit of pressure

PV Photovoltaic: technology that converts light into electrical power

pV Pressure vs volume

RPM Revolutions (or rotations) per minute

RTE Round-trip efficiency: the delivered energy divided by the input energy of an energy storage device starting and ending at a fully charged state

SCE Southern California Edison: a utility in California

SMLD Sterling Municipal Light Department

SOC State of charge: a measure of how full an energy storage is with respect to how much it can discharge

SPP Southwest Power Pool (see ISO)

TCO Total cost of ownership: a measure of the up-front (capex) and ongoing (opex) costs of an operating asset

TES Thermal energy storage: a means to store energy in the form of a temperature differential

tonne Unit of mass equal to 1,000 kgs

TVA Tennessee Valley Authority: an electric utility in the eastern part of the United States

TW Unit of power: equal to 1 trillion watts

TWh Terawatt-hour: unit of energy equal to 1 trillion Wh

UEP Urban Electric Power: a manufacturer of zinc-manganese dioxide batteries

UFLS Under frequency limit setting: a relay setting that will trip a breaker if the grid frequency falls below this value

UPS Uninterruptible power supply: a device to protect and power one or more loads from power disturbances.

USPTO United State Patent and Trademark Office

V Volt: a unit of voltage

VRB™ A trademarked name of a company that manufacturers VRFB

VRFB Vanadium-redox flow battery

W Watt. a unit of power: the rate of change of energy

Wh Watt-hour: a unit of energy equal to 3,600 Joules

XFMR Transformer: converts one ac voltage to another bidirectionally

ZBFB Zinc-bromine flow battery

About the Author

C. Michael Hoff received a bachelor of science degree in electrical engineering with a concentration in power from Drexel University in 1987, and a master of science degree in electrical engineering with a power concentration from Northeastern University in Boston, Massachusetts. He worked for American Power Conversion (APC) for 18 years, starting as a junior team contributor and working his way up to the director of research in the Smart-UPS Business Unit. There, he worked hands-on with fuel-cells, alternative battery chemistries, power conversion technologies, and communication protocols to advance the state-of-the-art of APC's UPS products.

In 2006, Mr. Hoff joined A123 Systems to help start the systems division, which became responsible for the development of EV and hybrid vehicle packs, vehicle starter, UPS and telcom, aircraft, and stationary energy storage batteries. There, he served in various roles as engineering manager, director of product safety and compliance, director of applications engineering, and director of research and technology. When NEC acquired the stationary storage division and created NEC Energy Solutions, Mr. Hoff became the chief technology officer and led a group to explore, evaluate, and model energy storage batteries and other technologies for use in current and future products. His team looked and evaluated all energy storage technologies as it related to their business model, including lithium-ion, non-lithium, flow-batteries, supercapacitors, and mechanical energy storage.

As NEC was winding down its operations in 2020, Mr. Hoff started HoffPower LLC to assist various battery developers, manufacturers, former customers, integrators, and developers that were optimizing their use of energy storage in their own businesses. Today, Mr. Hoff works full-time at American Battery Solutions developing products for grid energy storage.

Mr. Hoff is inventor and coinventor of more than a dozen patents and he has authored and coauthored several technical publications on batteries, power systems and topologies, and energy storage architectures and cost models.

Index

I

J

K

L

R

S

Artech House Power Engineering Library

Andres Carvallo, Series Editor

Advanced Technology for Smart Buildings, James Sinopoli

The Advanced Smart Grid: Edge Power Driving Sustainability, Second Edition, Andres Carvallo and John Cooper

Applications of Energy Harvesting Technologies in Buildings, Joseph W. Matiko and Stephen Beeby

Battery Management Systems, Volume I: Battery Modelings, Gregory L. Plett

Battery Management Systems, Volume II: Equivalent-Circuit Methods, Gregory L. Plett

Battery Management Systems for Large Lithium Ion Battery Packs, Davide Andrea

Battery Power Management for Portable Devices, Yevgen Barsukov and Jinrong Qian

Big Data Analytics for Connected Vehicles and Smart Cities, Bob McQueen

Design and Analysis of Large Lithium-Ion Battery Systems, Shriram Santhanagopalan, Kandler Smith, Jeremy Neubauer, Gi-Heon Kim, Matthew Keyser, and Ahmad Pesaran

Designing Control Loops for Linear and Switching Power Supplies: A Tutorial Guide, Christophe Basso

Electric Power System Fundamentals, Salvador Acha Daza

Electric Systems Operations: Evolving to the Modern Grid, Second Edition, Mani Vadari

Energy Harvesting for Autonomous Systems, Stephen Beeby and Neil White

Energy Storage Technologies and Applications, C. Michael Hoff

GIS for Enhanced Electric Utility Performance, Bill Meehan

IEC 61850 Demystified, Herbert Falk

Introduction to Power Electronics, Paul H. Chappell

Introduction to Power Utility Communications, Harvey Lehpamer

IoT Technical Challenges and Solutions, Arpan Pal and Balamuralidhar Purushothaman

Lithium-Ion Batteries and Applications: A Practical and Comprehensive Guide to Lithium-Ion Batteries and Arrays, from Toys to Towns, Volume 1, Batteries, Davide Andrea

Lithium-Ion Batteries and Applications: A Practical and Comprehensive Guide to Lithium-Ion Batteries and Arrays, from Toys to Towns, Volume 2, Applications, Davide Andrea

Lithium-Ion Battery Failures in Consumer Electronics, Ashish Arora, Sneha Arun Lele, Noshirwan Medora, and Shukri Souri

Microgrid Design and Operation: Toward Smart Energy in Cities, Federico Delfino, Renato Procopio, Mansueto Rossi, Stefano Bracco, Massimo Brignone, and Michela Robba

Plug-in Electric Vehicle Grid Integration, Islam Safak Bayram and Ali Tajer

Power Grid Resiliency for Adverse Conditions, Nicholas Abi-Samra

Power Line Communications in Practice, Xavier Carcelle

Power System State Estimation, Mukhtar Ahmad

Renewable Energy Technologies and Resources, Nader Anani

Signal Processing for RF Circuit Impairment Mitigation in Wireless Communications, Xinping Huang, Zhiwen Zhu, and Henry Leung

The Smart Grid as An Application Development Platform, George Koutitas and Stan McClellan

Smart Grid Redefined: Transformation of the Electric Utility, Mani Vadari

Sustainable Power, Autonomous Ships, and Cleaner Energy for Shipping, John Erik Hagen

Synergies for Sustainable Energy, Elvin Yüzügüllü

A Systems Approach to Lithium-Ion Battery Management, Phil Weicker

Telecommunication Networks for the Smart Grid, Alberto Sendin, Miguel A. Sanchez-Fornie, Iñigo Berganza, Javier Simon, and Iker Urrutia

A Whole-System Approach to High-Performance Green Buildings, David Strong and Victoria Burrows

For further information on these and other Artech House titles, including previously considered out-of-print books now available through our In-Print-Forever® (IPF®) program, contact:

Artech House
685 Canton Street
Norwood, MA 02062
Phone: 781-769-9750
Fax: 781-769-6334
e-mail: artech@artechhouse.com

Artech House
16 Sussex Street
London SW1V 4RW UK
Phone: +44 (0)20 7596-8750
Fax: +44 (0)20 7630-0166
e-mail: artech-uk@artechhouse.com

Find us on the World Wide Web at: www.artechhouse.com